生命急救技能

主编　许 虹　杨 勇　尉建锋　倪克锋

浙江科学技术出版社

图书在版编目(CIP)数据

生命急救技能/许虹等主编. —杭州：浙江科学技术出版社，2016.7

ISBN 978-7-5341-7250-2

Ⅰ.①生… Ⅱ.①许… Ⅲ.①自救互救—教材 Ⅳ.①X4

中国版本图书馆CIP数据核字(2016)第179591号

书　　名　生命急救技能

主　　编　许　虹　杨　勇　尉建峰　倪克锋

出版发行　浙江科学技术出版社

杭州市体育场路347号　邮政编码：310006

办公室电话：0571-85176593

销售部电话：0571-85176040

网　址：www.zkpress.com

E-mail：zkpress@zkpress.com

印　　刷　杭州杭新印务有限公司

经　　销　全国各地新华书店

开　　本　710×1000　1/16　　**印　张**　11.75

字　　数　190 000

版　　次　2016年7月第1版　　2016年7月第1次印刷

书　　号　ISBN 978-7-5341-7250-2　　**定　价**　35.00元

责任编辑　张祝娟　　**责任校对**　杜宇洁　马　融

责任美编　孙　菁　　**责任印务**　崔文红

编辑委员会

主　编　许　虹　杨　勇　尉建锋　倪克锋

副主编　楼　妍　应旭旻　赵桑桑　曹金科

编　者（以姓氏笔画排序）

刁文华　山东中医药高等专科学校

王　姗　杭州市萧山区第四中等职业学校

王晓蕾　杭州师范大学医学院

庄一渝　浙江大学附属邵逸夫医院

许　虹　杭州师范大学医学院

李冬梅　杭州师范大学医学院

李艳娟　杭州市中医院

杨　勇　杭州市中医院

杨　斌　杭州师范大学文化创意学院

吴育红　杭州师范大学医学院

应旭旻　杭州市急救中心

张丽君　杭州师范大学医学院

张祖勇　杭州市中医院

林子祺　杭州师范大学文化创意学院

赵桑桑　杭州卓健信息科技有限公司

胡文奕　杭州师范大学医学院

袁晓红　杭州师范大学医学院

晏慧敏　杭州师范大学医学院

钱　英　杭州师范大学医学院

倪克锋　杭州卓健信息科技有限公司

徐　鹏　杭州师范大学文化创意学院

陶月仙　杭州师范大学医学院

曹金科　杭州卓健信息科技有限公司

梁　琦　杭州师范大学医学院

尉建锋　浙江大学附属第一医院

董　锐　杭州师范大学医学院

鲁美丽　杭州市急救中心

蓝戈文　杭州卓健信息科技有限公司

楼　妍　杭州师范大学医学院

缪群芳　杭州师范大学医学院

秘　书　郭昕昕　杭州卓健信息科技有限公司

前 言

“江南忆，最忆是杭州”。杭州作为一座历史文化古城，也是一座创新之城，既充满浓郁的中华文化韵味，也拥有面向世界的宽广视野。2016年G20峰会在杭州举办，既是荣耀，更是机遇。

承办G20峰会是杭州向全世界展示中国公民素质的一个重要窗口，公民在面对突发事件时所呈现的反应能力是公民综合素质的核心内容之一。2016年1月14日，浙江省委常委、杭州市委书记赵一德赴杭州师范大学宣讲党的十八届五中全会精神，提出重点要做好服务保障G20峰会顺利举行。杭州师范大学作为杭州市唯一一所市属高校，服务杭州是其重要职能。杭州师范大学医学院护理系有近百年办学历史，其急危重症护理教学团队先后主编、出版了浙江省“十一五”重点规划教材、国家“十二五”和“十三五”规划教材《急救护理学》（包括双语）、国家“十二五”重点音像出版规划视听教材及指导用书等十余部书籍，主持浙江省省级精品课程《急危重症护理学》，获国家、省、市十余项教学成果奖。近十年，教学团队把课堂搬到部队、社区、广场、高校及全省养老护理、母婴护理人员的在职培训中，直接服务社会，以提升现场“第一目击者”的急救反应及施救能力，取得了一定的成效和不错的社会效应。基于前期的工作基础，急危重症护理学团队主动请缨，联合杭州市急救中心共同承接G20志愿者院前急救培训工作。

为了使接受培训的志愿者和民众尽快掌握院前急救技能，教学团队在前期多年教材编写的基础上，集临床专家、急救专家、移动健康专家、科普专家、美术团队的力量，倾力编写《生命急救技能》。该教材既有别于专业教材，又不同于一般的科普读物，本书将十七项常见突发状况应对措施，通过文字和漫画双重途径进行传递，通俗易懂，适合不同人群学习和应用。在体例编排上，以中文和英文双语形式呈现，与国际接轨；在风格设计上，彰显杭州风土人情和中国文化，将西湖新旧十景融入书稿，封面和内文以隶书、宋体、启功体等多种形式展现。《生命急救技能》可提升公民急救素质，既是专业知识走向民众的桥梁，又是杭州城市的一张名片，同时也让世界了解杭州，让杭州走向世界。

本书在编写过程中得到了多方的支持，感谢杭州市政府、杭州市卫生和计划生育委员会、杭州师范大学医学院、杭州市中医院、杭州师范大学文化创意学院和美术学院、浙江大学附属第一医院、杭州市急救中心、浙江大学附属邵逸夫医院的大力支持；感谢新西兰奥克兰医学院护理学院孙穗民老师审核英文稿；感谢杭州师范大学文创学院本科生、硕士研究生绘制急救技能的漫画。

尽管本书编写人员付出了辛勤的劳动，但由于时间仓促，难免有疏漏之处，希望广大同仁批评、指正，以便进一步修改和完善。

2016年7月

目录
CONTENT

苏堤春晓

CHAPTER

01 / 心肺复苏

Cardiopulmonary Resuscitation (CPR)

每个公民在遇到意外事件或急症发作、慢性病急性发作等危及生命的时候，如果第一时间没能获得紧急救助，不仅贻误急救的黄金时间，甚至会因此丧失宝贵的生命。挽救生命，其实每个人都能做到，关键时刻，你可成为挽救生命的第一施救者。请用你的手，给危难之人带去生的希望！

安卓用户请用
浏览器扫一扫打开

1.1 心肺复苏

❶ 判断

查看意识、判断脉搏：

1. 评估、判断现场是否安全（确保现场安全）。

2. 快速查看伤者有无意识，触摸颈动脉，判断有无脉搏（10秒钟内完成）。

❷ 求救

1. 确认伤者无意识反应后高声求救。

2. 拨打“120”。

3. 若现场附近有自动体外除颤仪(automated external defibrillator, AED)，设法获取。

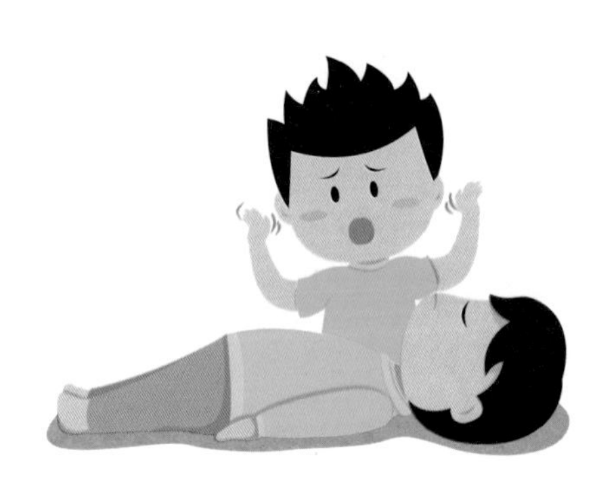

安置体位

（1）去枕仰卧位，确认硬板床或地面。

（2）头、颈、躯干在同一轴线。

（3）双手放于两侧，身体无扭曲。

❸ 按压

1. 解开衣领、裤带，暴露伤者胸腹部。

2. 选择按压部位：两乳头连线与胸骨交界处（胸骨中下1/3交界处）。

3. 双肘伸直，两手相扣，用掌根。

4. 紧贴胸部垂直按压。

5. 每分钟按100—120下，按压深度5—6厘米。

6. 保证每次按压后胸部回弹。

7. 尽可能减少按压中断（<10秒钟）。

按压（30次） ⇄ 循环 ⇄ 人工呼吸不压（2次）

④ 打开呼吸道

1. 口中若有异物先取出，保持呼吸道通畅。

2. 一手按压额头使头部后仰，另一手提起下巴（压额举颏法）。

⑤ 人工呼吸

连续吹两口气，每口气1秒钟（吹气可见胸廓起伏）：

1. 捏紧伤者鼻子做人工呼吸。

2. 使用防护设备。

⑥ 除颤

1. 打开电击器。

2. 依照说明贴覆导片。

3. 插上导线连接除颤仪。

4. 依照指示执行电击。

1.2 除颤仪的使用

AED使用

打开AED

↓

清洁皮肤，选择导联Pads

↓

粘贴多功能电极片，接上多功能电缆

↓

按下自动分析键，听指示操作

↓

仪器自动充电

↓

充电完毕，所有人离开，按下电击键，电击完成

↓

若不需除颤，转到监护状态

准备	判断伤者状况，心电监护提示“室颤”需紧急电除颤，呼叫帮忙
	去枕平卧，开放气道，暴露前胸
评估	迅速擦干伤者胸部皮肤，迅速检查除颤仪后报“性能良好”
	观察周围环境，确保安全
开机选模式	正确开启除颤仪，选择模式（室颤一非同步模式），再次确认发生室颤
涂导电糊	在电极板上涂以适量的导电糊

选能量	选择正确除颤能量（单相波成人200焦耳，双相波成人120—200焦耳，儿童2—4焦耳/千克）
充　电	开始充电
	请“旁人离开”
安放电极板，紧贴皮肤	电极板位置安放正确，一个电极板置于右锁骨下胸骨右侧，另一电极板放在左乳头的左下方（心尖部），两电极板之间的距离不得小于10厘米
	电极板压力适当，不得歪斜
与伤者保持安全距离	操作者身体不能与伤者接触
	除颤前确定周围人员没有直接或间接与伤者接触
放　电	充电并显示可以除颤时，双手拇指同时按压放电按钮电击除颤
除颤结束	除颤结束移开电极板，观察是否恢复，如恢复窦性心律，报告“心跳恢复窦性”，将能量开关恢复至零位
	清洁皮肤，安置伤者（协助伤者整理衣物，取舒适卧位）
	清洁除颤电极板，正确归位电极板（如心律未恢复，继续胸外按压5个周期后再次除颤）
	密切监测伤者病情，洗手并记录情况

Cardiopulmonary Resuscitation (CPR)

Cardiopulmonary resuscitation (CPR) is a lifesaving technique useful in many emergencies, including heart attack or near drowning, in which someone's breathing or heartbeat has stopped. Time is life. Every citizen could be the first person to respond. In life-threatening conditions, you are the protector of lives.

1. Verify scene safety and check for responsiveness

(1) Verify scene safety, check if there are any potential hazards that exist to either the rescuers or the victim and even the bystanders (ensure the scene is safe).
(2) Check for responsiveness. Tap the victim's shoulder and shout, "Are you OK?"

2. Get nearby help

(1) If the patient is not responsive, shout for nearby help.
(2) If someone else is available, send that person to dial 120 or other emergency number, and get AED.

3. Assess for breathing and pulse

(1) Assess the victim for normal breathing and pulse at the same time.
(2) If you do not definitely feel a pulse within 10 seconds, and the victim is not breathing or only gasping, start CPR immediately, begin with chest compressions.

4. Perform high-quality CPR

(1) Positioning.

① Place the victim on his or her back, and lie on a firm surface.

② Put the victim's arms beside the body, and keep the victim's head, neck and back in a line.

(2) Compression (C).

① Remove clothing from the chest area.

② Position of compression: the center of the chest or the lower half of the breastbone (sternum) is the compression area.

③ Place the heel of one hand on the lower half of the sternum. Place the other hand on top of the first, palms-down. Fingers should be interlaced and should be kept off the chest. Position your body so your shoulders are directly over your hands. Straighten your arms and lock your elbows.

④ Press down with both hands directly over the breastbone to perform a compression.

⑤ Press down at 5 to 6 cm , with a rate of 100 to120 per minute.

⑥ After each compression, release all pressure on the chest to allow the chest to recoil completely.

⑦ Minimize interruptions in compressions (<10s).

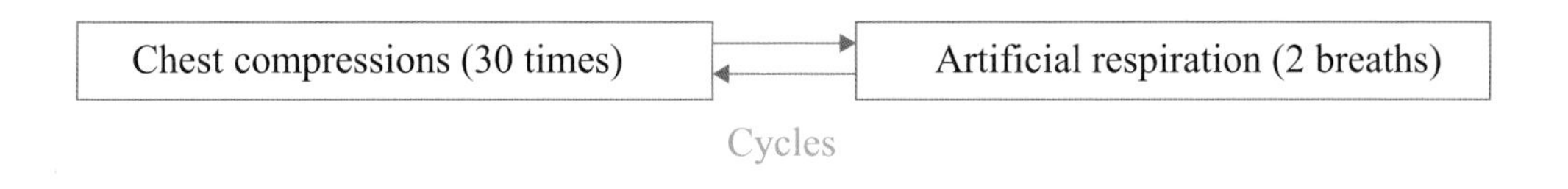

Cycles

5. Airway Opening (A)

(1) If there is any object that can be easily removed in the mouth, remove it.

(2) Place one hand on the victim's forehead and push with your palm to tilt the head back. Place the fingers of the other hand under the bony part of the lower jaw near the chin (with head tilt-chin lift).

6. Rescue breaths with mouth-to-mouth breathing (B)

(1) The rescuer should pinch the victim's nose with thumb and forefinger, make a mouth-to-mouth seal, give two breaths, and each full breath should take one second. Blow steadily until the chest rises.

(2) Use protective equipments.

(3) Repeat 30 chest compressions following two rescue breaths (Compression to Ventilation ratio is 30: 2).

(4) If the AED is available, apply it and follow the prompts.

(5) Continue CPR until you see signs of life or until medical personnel arrive.

7. The application of AED

(1) Turn on the AED，follow the AED prompts as a guide to next steps.

(2) Clean and dry the skin, attach the adhesive AED pads to the victim's bare chest. Attach the AED connecting cable to the AED device.

(3) Clear the victim. Press the analyze button, and allow the AED to analyze the rhythm.

(4) If the AED advise a shock, it will tell you to clear the victim and will start charge automatically. Once charged, check to make sure no one is touching the victim, press the button for electric shock following the advice on the AED.

(5) If no shock is advised or after shock delivery, immediately resume CPR, start with chest compressions.

(6) After 2 minutes of CPR, the AED will prompt you to clear the victim and analyze the rhythm (repeat step 3 and 4).

曲院风荷

CHAPTER

02 / 止血

Hemostasis

出血是创伤后最常见、也是最严重的并发症之一。失血量达到总血量20%时即会出现明显症状和体征；失血量达到总血量40%时，就有生命危险。因此，为减少创伤的死亡率和伤残率，必须争分夺秒采取有效的止血措施。

2.1 出血的分类

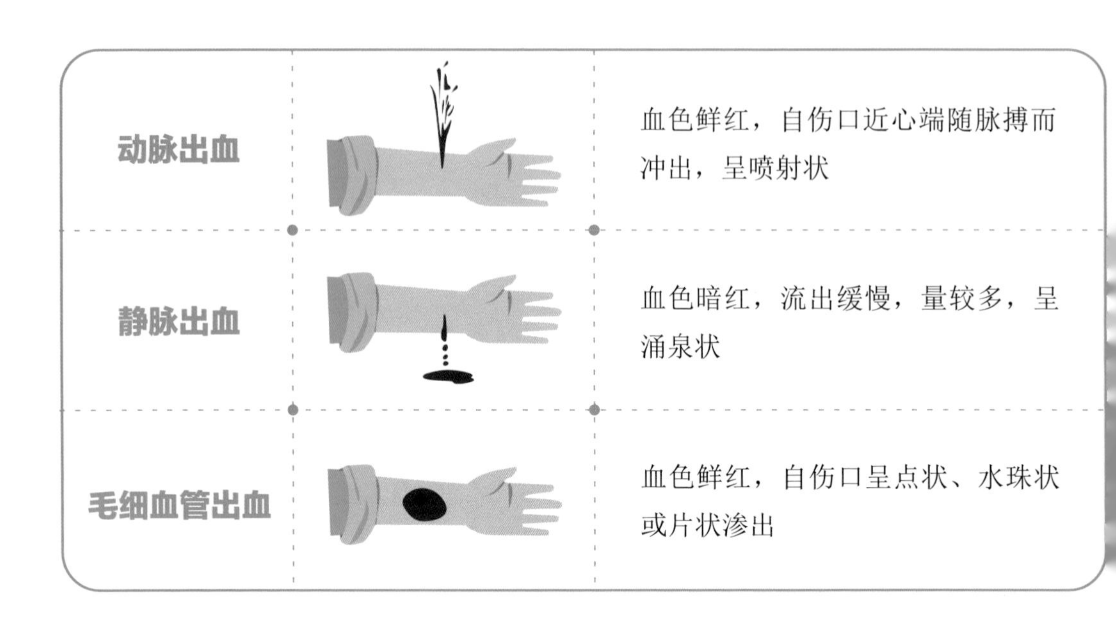

	描述
动脉出血	血色鲜红，自伤口近心端随脉搏而冲出，呈喷射状
静脉出血	血色暗红，流出缓慢，量较多，呈涌泉状
毛细血管出血	血色鲜红，自伤口呈点状、水珠状或片状渗出

2.2 止血物品

现场紧急情况下，各种有弹性的纺织物或者止血带都可用于紧急止血，如消毒敷料、绷带，甚至干净的毛巾、布料、围巾等，还有充气止血带、橡胶止血带等。

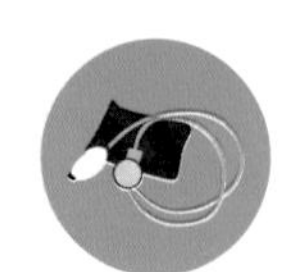

禁止使用电线或铁丝等坚硬无弹性物品。

2.3 止血方法

2.3.1 指压止血法

指压止血法适用于头、面、颈部和四肢的外出血，主要根据动脉走行位置，用手指压迫伤口近心端的动脉。

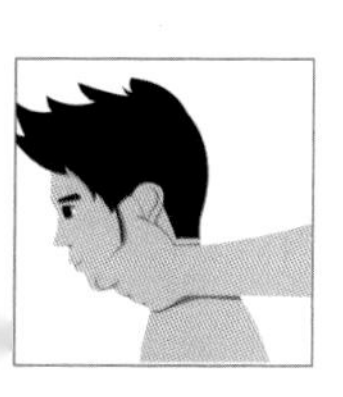

头顶部及前额出血

同侧耳屏上前方1.5厘米处

压迫颞浅动脉

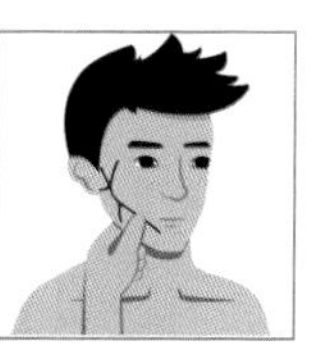

面部出血

下颌骨下缘、咬肌前缘凹陷

压迫同侧面动脉处搏动点

枕部出血

耳后乳突下稍往后的搏动点

压迫同侧枕动脉

颈部、面部、头皮部出血

气管外侧与胸锁乳突肌前缘中点之间

压迫同侧颈总动脉

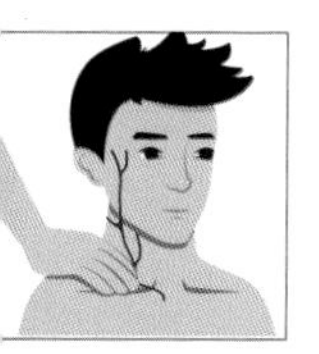

肩部、腋部、上臂出血

锁骨上窝中部，胸锁乳突肌外缘的搏动点

压迫同侧锁骨下动脉

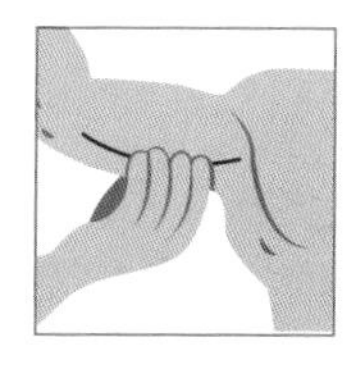

前臂出血

肱二头肌内侧沟中部搏动点

压迫同侧肱动脉

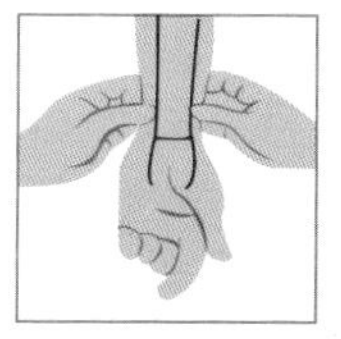

手掌、手背出血

手腕横纹稍上方的内、外侧搏动点

压迫同侧尺、桡动脉

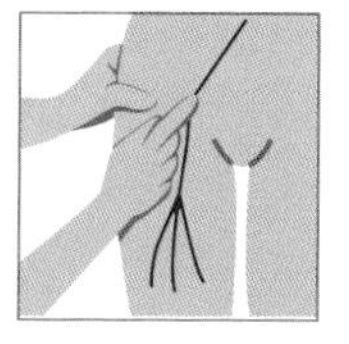

大腿出血

大腿根部腹股沟中点偏内侧的下方搏动点

压迫同侧股动脉

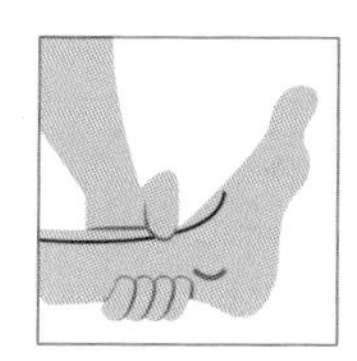

足部出血

足背中部近脚腕处和足跟与内踝之间的搏动点

压迫同侧的胫前、胫后动脉

压迫止血会影响肢体组织的血液供应，应限制使用，每次施压时间不能超过10分钟。

2.3.2 加压包扎止血法

加压包扎止血法是最常用的止血方法，既可止血，又可包扎伤口。

适用于 小动脉，中、小静脉和毛细血管出血。

方 法 先用无菌敷料将伤口覆盖，再用纱布、绷带作适当加压包扎，松紧度以能达到止血为宜，必要时可将手掌放在敷料上均匀加压。

2.3.3 填塞止血法

适用于 适用于大腿根、腋窝、肩部等难以用一般加压包扎所处理的较大而深的伤口出血。

方 法 最好用无菌敷料填入伤口内，压住破裂血管，外加大块敷料加压包扎。

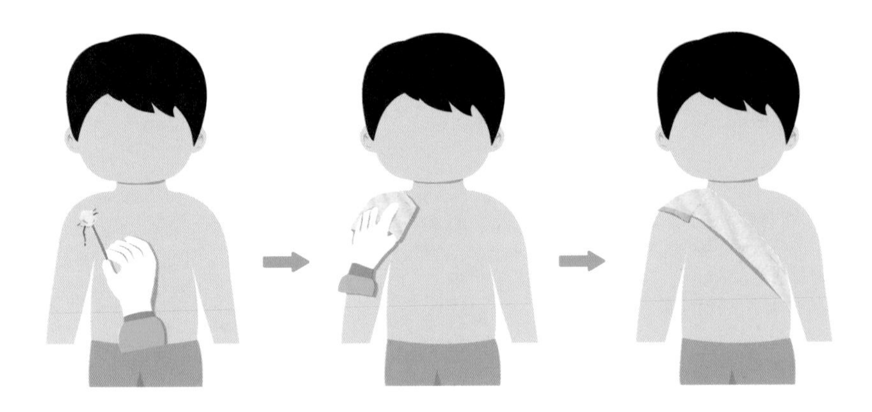

2.3.4 止血带止血法

适用于 四肢较大动脉的出血，或采用加压包扎后不能有效控制的大出血。常用的有充气止血带和橡皮止血带两种，在紧急情况下也可用绷带、布带、三角巾等代替。

止血带是止血的应急措施，在使用过程中包扎过紧会压迫损害神经或软组织，过松则起不到止血作用，时间过久则会引起肌肉坏死、厌氧细菌感染，甚至危及生命。

注意事项

- **部位要准确**

  止血带应扎在伤口的近心端，尽量靠近伤口。

- **皮肤与止血带之间要加衬垫**

  这样可以避免损伤皮肤，切忌用绳索或铁丝直接加压。

- **压力要适度**

  这样可以避免损伤皮肤，切忌用绳索或铁丝直接加压。

- **凡是上止血带的伤者必须做标记**

  标明其日期、时间和部位。

- **定时放松**

  原则上止血带限于1小时左右，如为充气式止血带也不宜超过3小时，且应每30—60分钟放松一次，放松时间视出血情况而定。

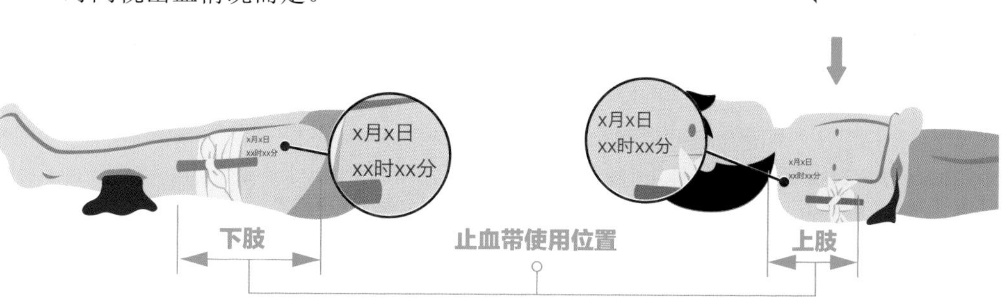

Hemostasis

Bleeding is one of the most common symptoms of any injuries. Massive amounts of bleeding could be vital and requires urgent intervention.

1. Three main types of bleeding

(1) Arterial bleeding: The blood is bright red, coming out in large volume and in spurts with the heartbeat.

(2) Venous bleeding: The blood is darker red, flowing slowly and steadily.

(3) Capillary bleeding: The blood flows very slow and small in quantity.

2. Equipment

In the event of an emergency, all kinds of elastic fabric or tourniquets can be used for controlling external bleeding, but do not use any hard and inelastic goods, such as an electric wire or iron wire.

3. Methods to control bleeding

(1) Pressure point compression.

Apply direct pressure to an artery. This method can be effective in stopping external bleeding in the head, face, neck, arms and legs, with pressure on the specific major arteries in the body. Push the artery against bone, and press down firmly on the artery between the bleeding site and the heart.

(2) Apply direct pressure.

This method is usually effective for controlling bleeding from an arteriole, medium-sized vein, veinlet, and capillary. Elevate the wound above the heart, cover the wound with a sterile dressing pad or gauze, bandage or other clean available cloths, such as T-shirt, then apply firm pressure on the wound. If blood soaks through the pad, do not remove the pad, and add another bandage on the top of it, then replace the top pad.

(3) Packing hemostasis.

It is effective for stopping the bleeding from severe and deep wounds at the end of the leg, armpit, shoulder, etc. Fill in the wounds with sterile dressings, press the broken blood vessels, and apply the pressure bandage to the wounds with a large dressing.

(4) Tourniquet.

A tourniquet is used in controlling severe bleeding from limbs. Tourniquets are emergency measures for trained people to use; attention should be paid to prevent a secondary compression injury.

Caution for using tourniquet

(1) The position must be accurate: tourniquet should be placed on the proximal to the wounds (between the open wound and the heart), and close to the wound but not directly over the wound.

(2) Place some padding underneath the tourniquet.

(3) The pressure should be moderate, with the distal artery pulse just disappearing as the tourniquet is bonded.

(4) The tourniquet should be marked with the date, time and location.

(5) Release the tourniquet at intervals, at every 30~60minutes, loosen it once, and the time you loosen it for depends on the condition of the bleeding.

南屏晚钟

CHAPTER

03 / 包扎

Bandaging Techniques

包扎的目的在于保护伤口，减少伤口感染和再损伤；局部加压，帮助止血；固定伤口上的敷料、夹板，撑托受伤的肢体，可使伤部舒适安全，减轻痛苦。

3.1 用物

特制材料 绷带、三角巾、四头带、多头带、丁字带等。
就便材料 洁净的毛巾、被单、丝巾、衣物等。

3.2 包扎的基本方法

3.2.1 绷带基本包扎法

绷带基本包扎法是用途最广、最方便的包扎方法，常用的基本方法有6种，根据包扎部位形状的不同而采用合适的方法。

1 螺旋形包扎法

将绷带环行缠绕数圈后，再斜行向上缠绕，每圈遮盖上一圈的1/3—1/2部位。

适用于 包扎直径基本相同的部位如上臂、躯干、大腿等。

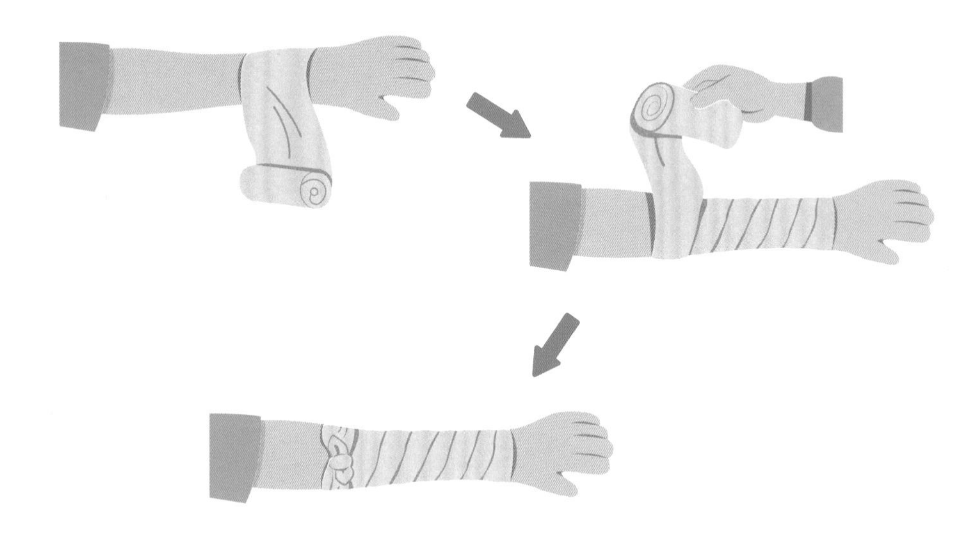

❷ 螺旋反折包扎法

每圈均把绷带向下反折，遮盖上一圈的1/3—1/2，反折部位应相同，使之成一直线。

适用于 包扎直径大小不等的部位，如前臂、小腿等。

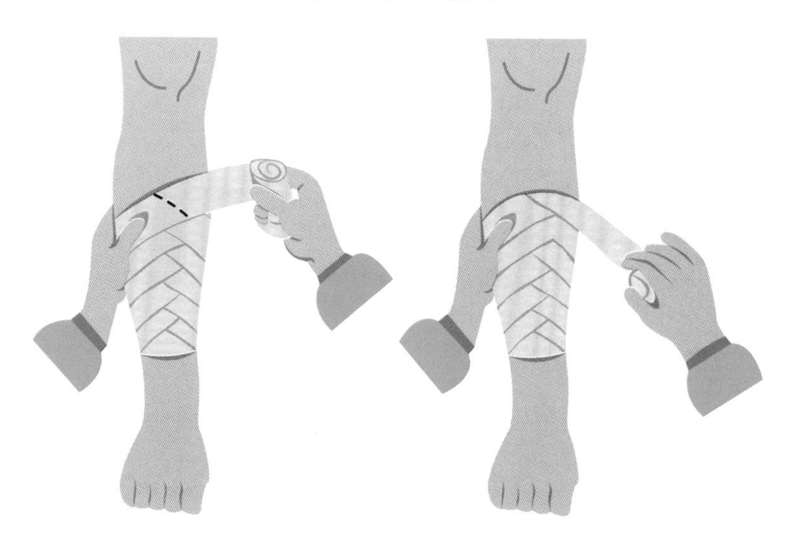

❸ 环形包扎法

将绷带环形重叠缠绕，下一圈必须遮盖上一圈，结束时用胶布固定尾端或将带尾分成两头，并在此处打结固定。

适用于 绷带包扎的开始与结束，或包扎粗细相等部位的小伤口，如颈、腕、胸、腹等处。

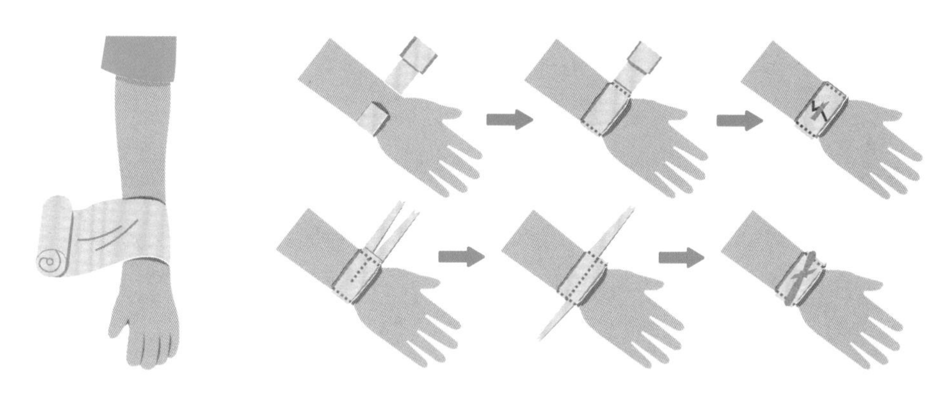

❹ 蛇形包扎

将绷带以环形包扎法缠绕数圈后，以绷带宽度为间隔，斜行上缠，每圈互不遮盖。

适用于 维持敷料或夹板固定。

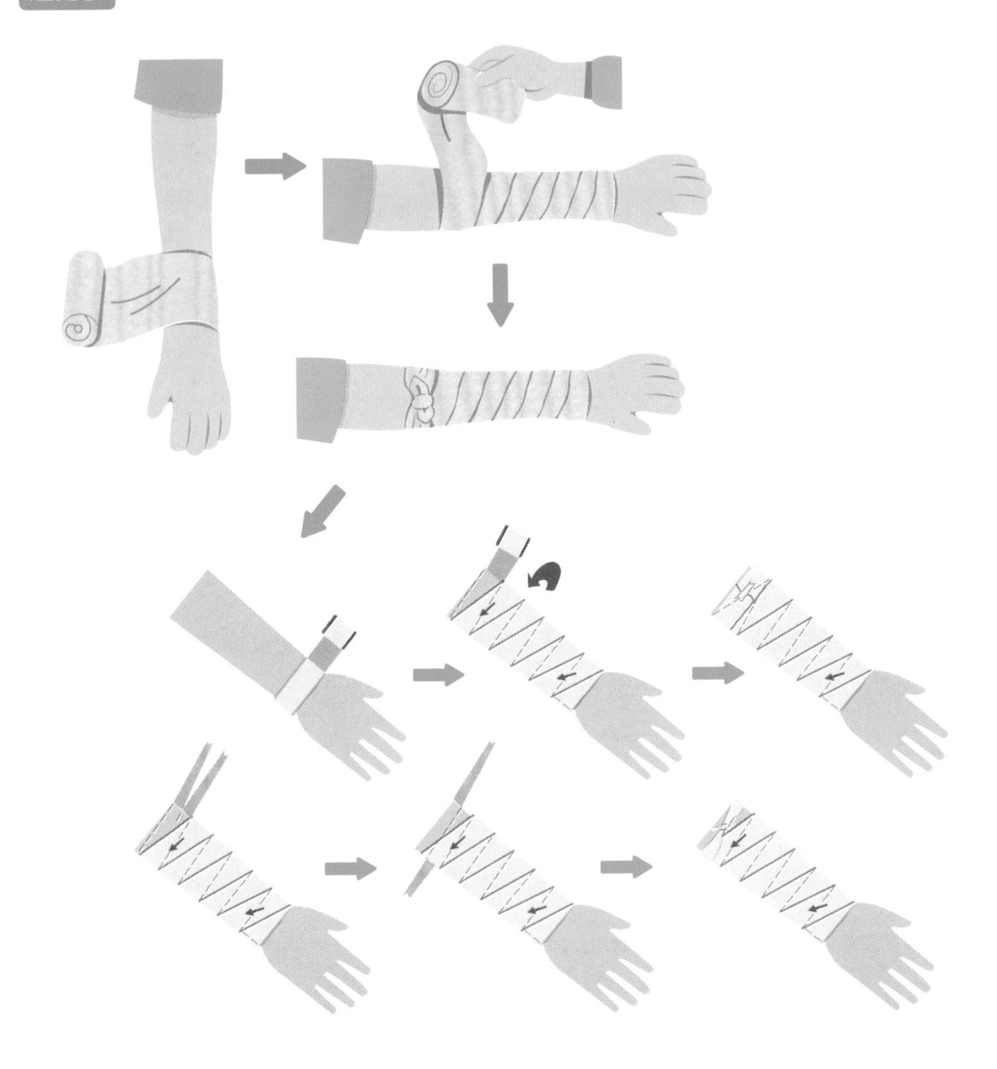

❺ “8”字形包扎

在伤处上下，将绷带由下而上，再由上而下，互相交叉包扎重复作“8”字形旋转缠绕，每圈遮盖上一圈的1/3—1/2 。

适用于 包扎屈曲的关节，如肩、肘、髋、膝关节等部位。

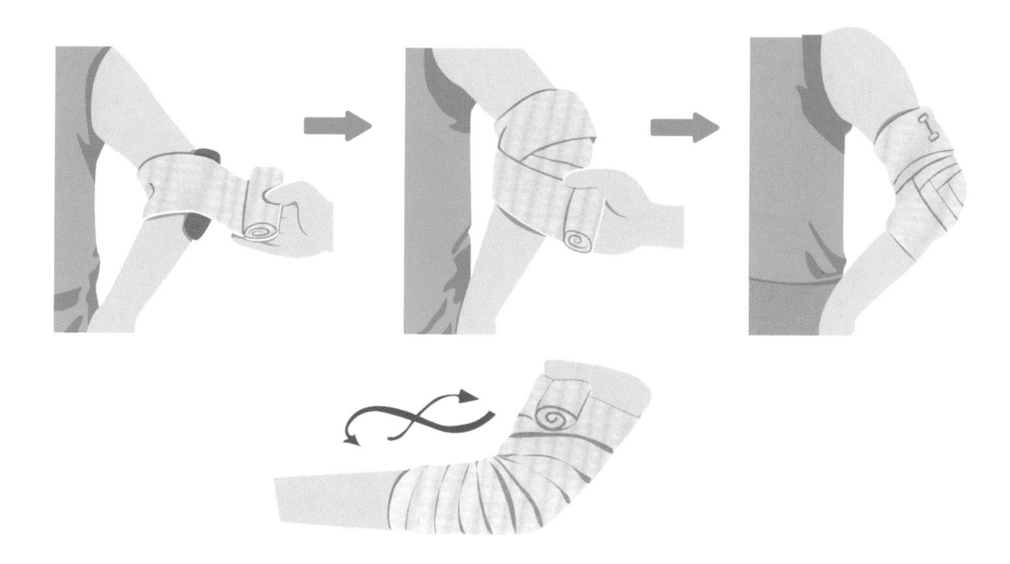

❻ 回返包扎

以环形包扎法缠绕数圈后，在中央来回反折，一直到该端全部包扎后，再作环形固定。

适用于 包扎有顶端的部位如头部、断肢残端。

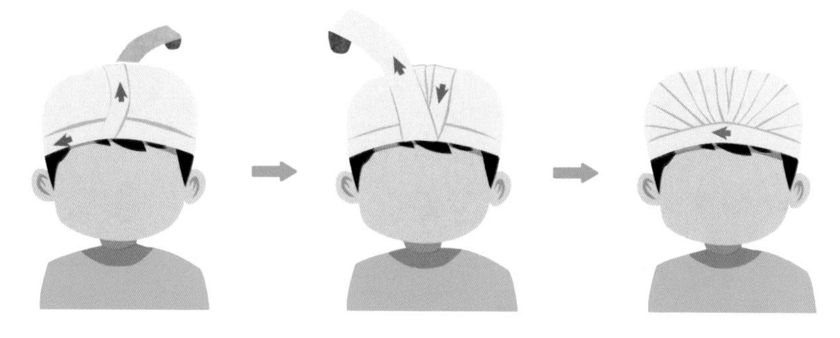

3.2.2 三角巾包扎法

三角巾可灵活运用于身体各部位较大伤口的包扎。使用时可根据需要折叠成不同形状，将三角巾顶角偏左或偏右的位置折到底边中点，可将三角巾折叠成燕尾形，根据包扎部位的不同调整燕尾巾夹角大小，也可折叠成带状作为悬吊带或用作肢体创伤及头、眼、膝、肘、手部较小伤口的包扎。

1 头顶帽式包扎法

（1）将三角巾底边向上反折约3厘米。

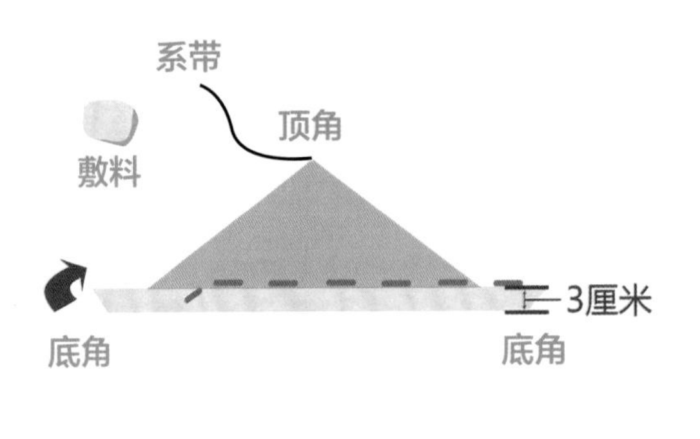

（2）将折缘朝内放在前额，与眉齐平。

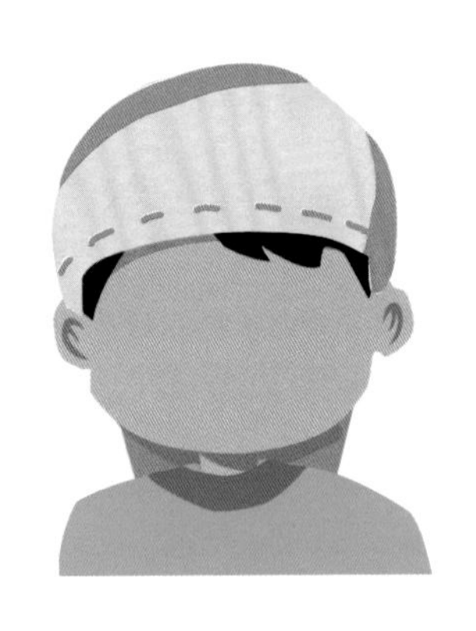

（3）将顶角越过头顶，拉向头后，两底角从两耳上方绕至枕后交叉。

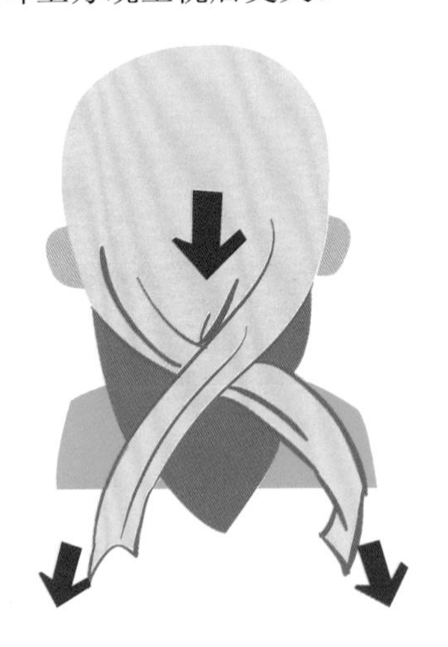

（4）将底角绕回前额打结固定。

（5）将顶角拉紧向上反折平整塞入头后部交叉处内。

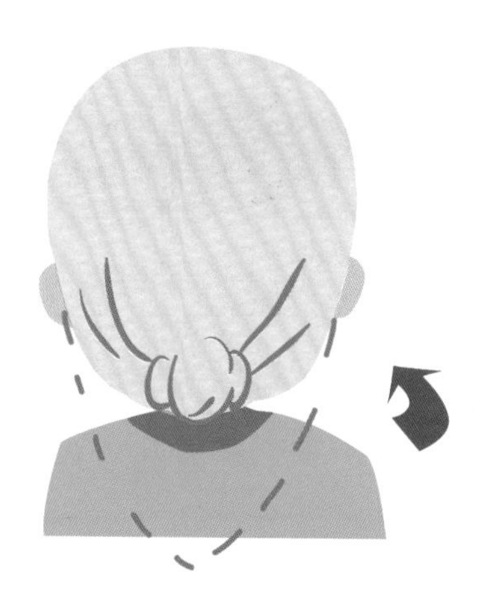

❷ 单肩燕尾巾包扎

（1）将三角巾折成燕尾式。

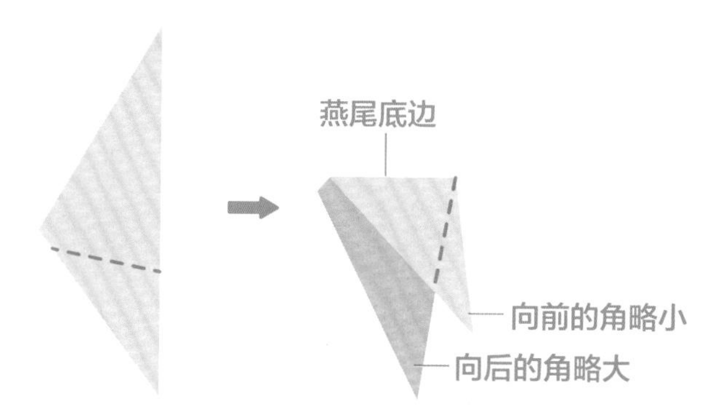

（2）将燕尾巾夹角朝上，放在伤侧肩上。

（3）向后的一角略大并压住向前的角，燕尾底边包绕上臂上部打结，然后两燕尾角分别经胸、背拉到对侧腋下打结。

❸ 单肩包扎法

把三角巾一底角斜放在胸前对侧腋下，将三角巾顶角盖住后肩部，用顶角系带在上臂三角肌处固定，再把另一个底角上翻后拉，在腋下两角打结。

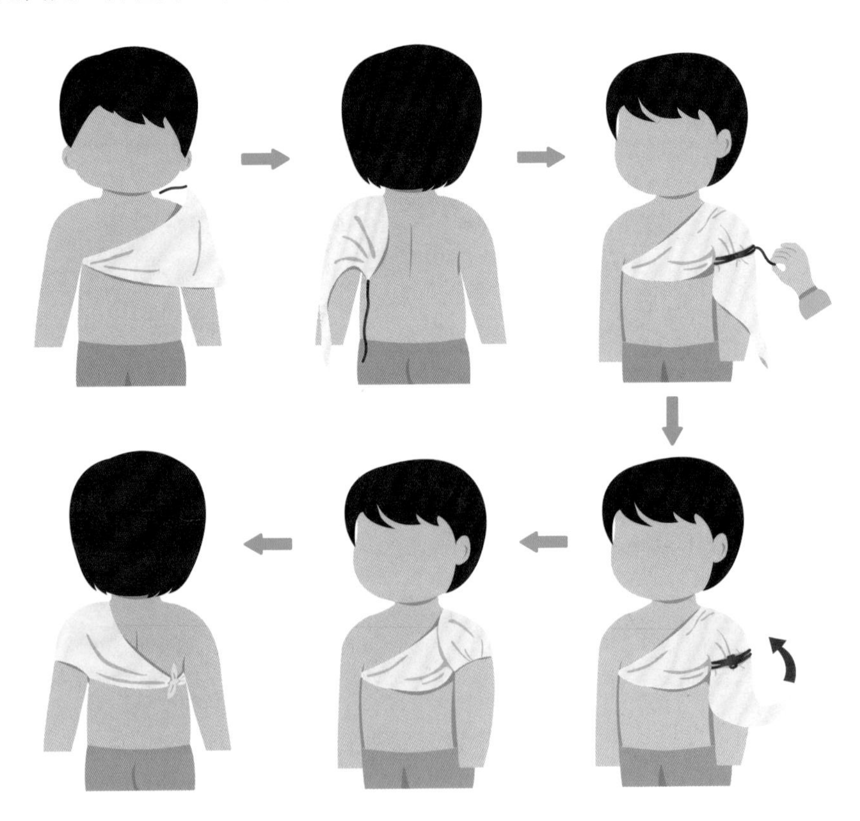

❹ 双肩燕尾巾包扎

（1）将三角巾折成燕尾式，两燕尾角等大。

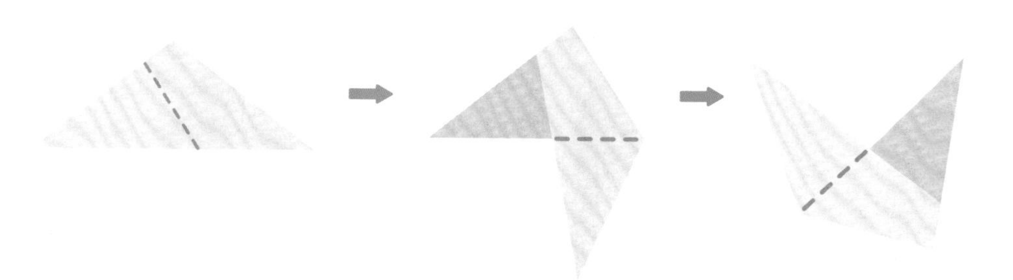

（2）夹角朝上对准颈部，燕尾披在双肩上。

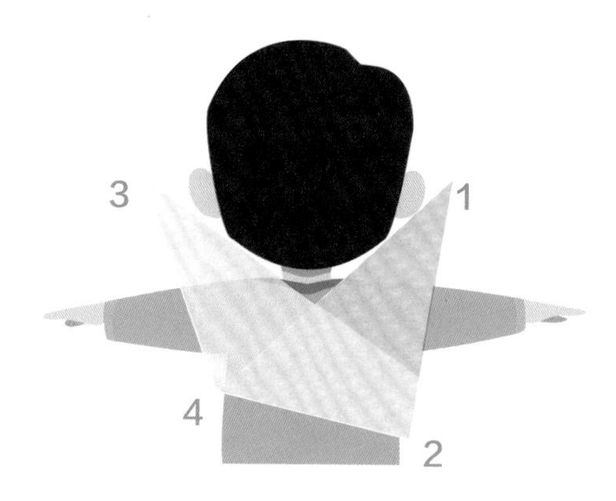

（3）两燕尾角分别包绕两肩膀拉到腋下打结。

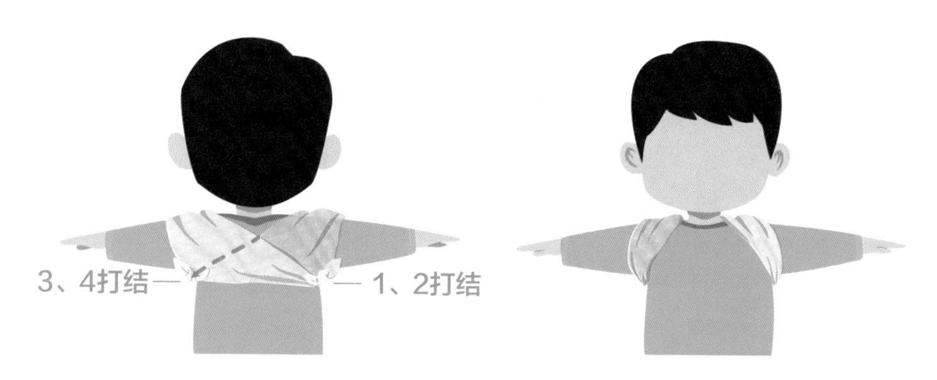

⑤ 胸部包扎法

将三角巾底边横放在胸部，约在肘弯上3厘米处，顶角越过伤侧肩，垂向背部，三角巾的中部盖在胸部的伤处，两端拉向背部打结，顶角也和此结一起打结。

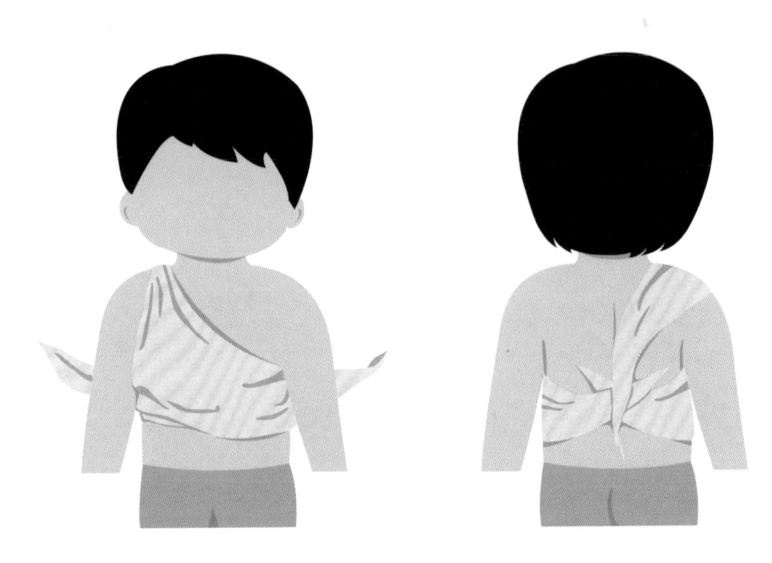

⑥ 背部包扎法

（1）方法与胸部相同，只是位置相反，结打于胸部。将三角巾底边横放在背部，约在肘弯上3厘米处，顶角越过伤侧背，垂向胸前。

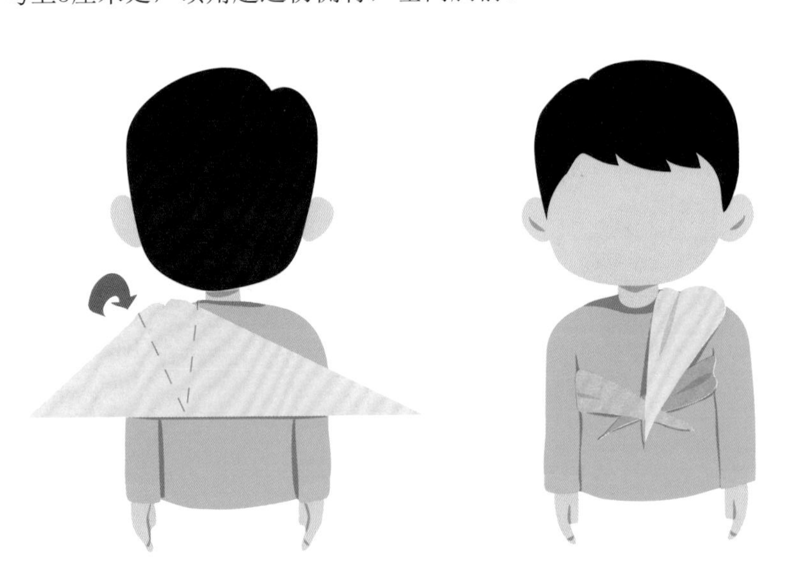

（2）三角巾的中部盖在背部的伤处，两端拉向胸前打结，顶角也和此结一起打结。

❼ 臀部包扎法

（1）将三角巾顶角朝下，底边横放于脐部并外翻10厘米左右。

（2）拉紧两底角至腰背部打结，顶角经会阴拉至臀上方，与底角余头打结。

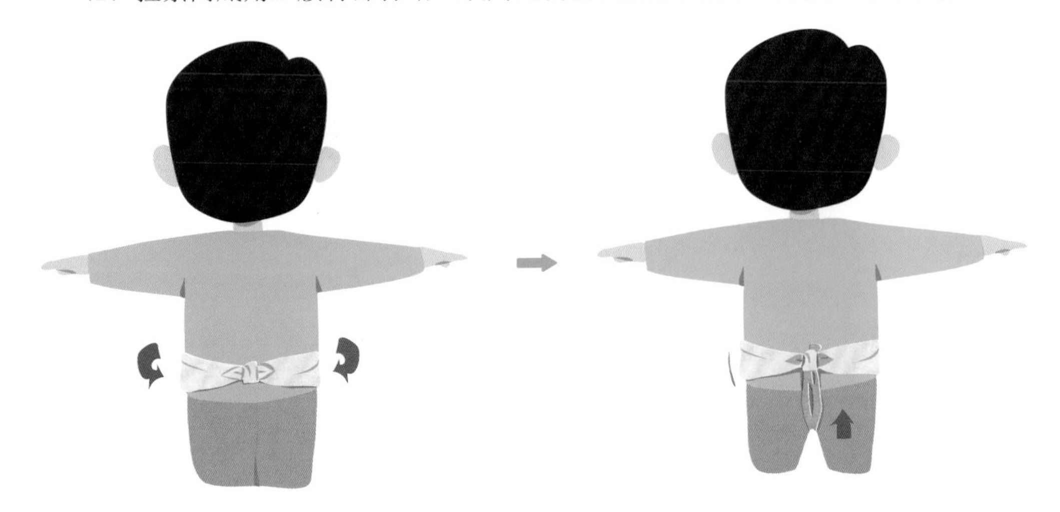

❽ 手、足包扎法

将手或足放在三角巾中央，指（趾）尖对着顶角，底边位于腕部，将顶角提起反盖于全手或足背上，拉左右两底角交叉压住顶角，绕回腕（踝）部，于掌侧或背侧打结固定。

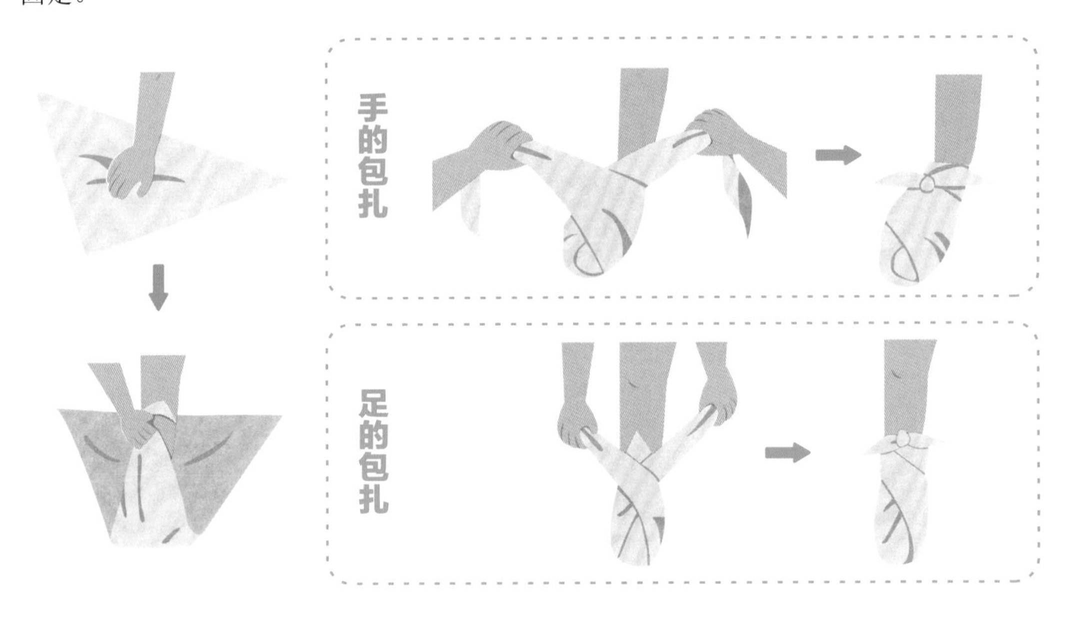

❾ 上肢包扎法

（1）将三角巾一底角打结后套在伤侧手上，打结余头留长些备用。

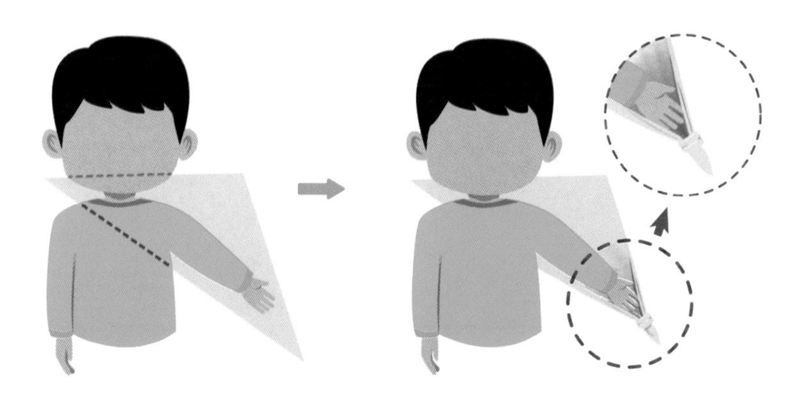

（2）另一底角沿手臂后侧拉到对侧肩上，顶角包裹伤肢适当固定。

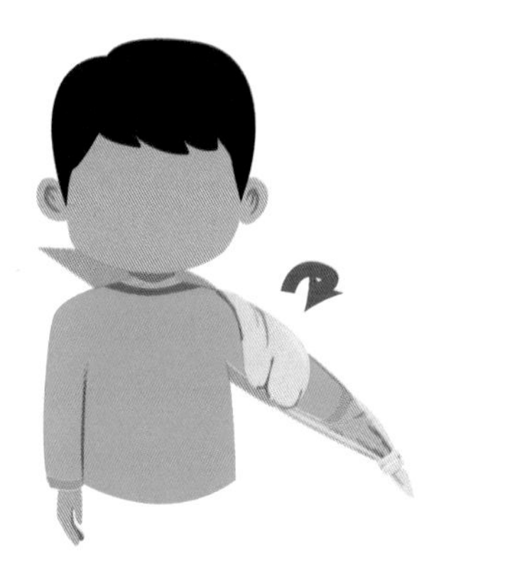

（3）前臂屈到胸前，拉紧两底角打结。

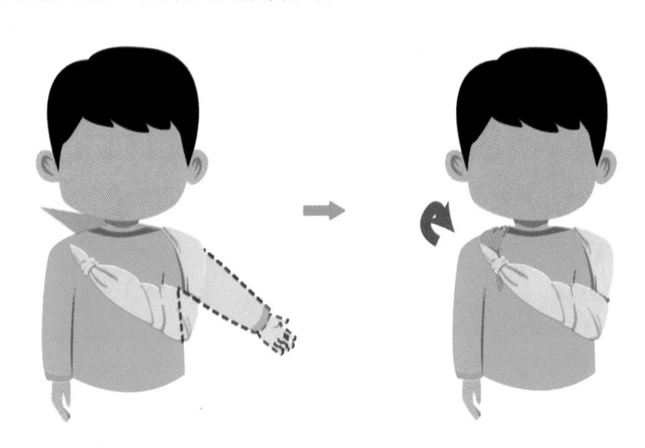

Bandaging Techniques

Wound dressings are important measures to protect the wound, reduce infection and prevent any further injury.

1. Materials

Different types of bandages are used for wound dressing. In an emergency, you can use any clean, absorbent material as a bandage.

2. Methods of bandage application

There are six ways of bandaging, which are applied to different parts of the body.

(1) Spiral bandage: It is usually used for cylindrical parts of the body, such as the limbs, the torso, legs. The spiral begins distally on an extremity and wind proximally. Start with the circular method with one or two complete overlaps, then a half or two-thirds width of the preceding turn is covered by the next turn. Move evenly up the extremity to provide even support.

(2) Reverse spiral bandage: It is a spiral bandage where the bandage is folded back on itself by 180° after each turn. Like the spiral bandage, part of the preceding turn is covered by 1/3-1/2 of the width of the bandage. It is used for bandaging forearm and calf.

(3) Roller (circular) bandage: It is used for the beginning and the end of other bandage patterns, or for covering small wounds on the extremity with an equal-circumference. The upper layer should cover up the lower layer.

(4) Snake-shaped bandage: It is suitable for fixing dressing or splint. Anchor the bandage for several rounds, then encircle obliquely at bandage width intervals, with each layer is not overlapped.

(5) Figure-of-eight turns bandage (Elbow and knee bandages): It is used to bind a flexing joint or body part below and above a joint. There are five steps: ① Flex the joint slightly. Place the tail of the bandage on the inner side of the joint, pass the bandage over and around to the outside of the joint, and make one-and-a-half turns; ② Then pass the bandage to the inner side of the limb, and make a turn around the limb, covering the upper half of the bandage from the first turn; ③ Then pass the bandage from the inner side of the upper limb to just below the joint, and make one diagonal turn below the joint to cover the lower half of the bandaging from the first straight turn; ④ Continue to bandage diagonally above and below the joint in a figure-of-eight. Repeat this figure-of-eight maneuver as many times as necessary to fix the dressing properly, and overlap each layer by one thirds to half the width of the bandage; ⑤ Finish bandaging with two straight turns around the limb and secure the end of the bandage.

(6) Recurrent bandage: It is suitable for blunt body parts (the head or a limb's stump). Begin with one circular turn, and then the bandage is applied repeatedly from one side across the top to the other side of the blunt body part. To be able to fix the recurrent turns well, the entire length of the blunt body part should be covered. Recurrent bandages are fixed using circular turns.

3. Triangular bandage

A triangular bandage or cravat is a large triangle of cloth, often a loose-weave cotton cloth. It has flexible function in wound dressings and can be folded into different shapes according to the condition. Swallow tail and ribbon shapes are commonly used.

(1) Triangular bandage for the head.

① Fold back the bandage base three centimeters to make a hem. Place the middle of the triangle base on the forehead, just above the eyebrows, with the hem on the outside.

② Let the point fall over the head and down the back of the head. Bring the ends of the triangle around the back of the head above the ears, then cross them over the point.

③ Carry the base around to the forehead, and tie a knot.

④ Bring the point up then tuck it over and in the bandage where it crosses the back part of the head.

(2) Triangular bandage for a shoulder.

① Fold the point of the triangular bandage to the middle of the base and then fold lengthwise to the desired width, like the swallow tail.

② Place the point of cravat over the injured shoulder near the neck.

③ Bring the cravat's ends under the opposite armpit and tie it slightly in front of or behind it.

(3) Triangular bandage for both shoulders.

① A cravat bandage is made by bringing the point of the triangular bandage to the middle of the base and then folding lengthwise to the desired width.

② Make the cravat point towards the neck, place it on both shoulders.

③ Tie a knot under the armpits after dressing shoulders with the other two edges respectively.

(4) Triangular bandage for the chest.

① Place the point of the bandage on the shoulder of injured side.

② Bring bandage down over chest to cover the injured chest. The middle of the base of the bandage is directly below the wound.

③ Carry two ends of the base around body and tie a square knot.

④ Bring the point of the bandage down and tie to one of the ends of the first knot.

(5) Triangular bandage for the back.

Use the same methods as bandage for chest but with an opposite position. Then tie a knot on the chest.

(6) Bandage for hips.

① Make the upper corner point downwards, place the bottom on the navel and overlap it about 10 centimeters outwards.

② Bring the two points to the back of the waist and tie a knot.

③ Bring the upper point to the top of the hip across the perineum, and tie a knot with the bottom points.

(7) Triangular bandage for hands and feet.

Lay the bandage flat. Place the hand or foot on the center of the triangular bandage with the fingers toward the point. Fold the point over the hand or foot. Cross the ends of the bandage over the hand or foot, and then pass the ends around the wrist or ankle in opposite directions. Tie the ends in a square knot.

(8) Bandage for the upper arm.

① Tie a knot on the base point of the triangular bandage and put the injured limbs in.

② Place another base along the back of the arm to the contralateral shoulders. Fix the position of the injured limbs with the upper corner.

③ Bend the injured arms across the chest and tie ends into a knot.

满陇桂雨

CHAPTER

04 / 固定

Fixation

对于骨折伤者，急救时应将骨折部位进行临时固定，其目的是限制骨折断端活动，防止骨折周围组织的继发性损伤，减轻疼痛和便于搬运。

4.1 用物

特制材料 夹板，有木质夹板、铁丝夹板、塑料制品夹板和充气夹板等。

就便材料 竹板、椅子、木棒等。紧急情况下可使用木棒等长条形硬物，甚至可用健侧肢体。

4.2 常用固定方法

❶ 上臂骨折

将夹板放于伤臂外侧，在骨折部位上下两端固定，将肘关节屈曲90度，使前臂呈中立位，再用三角巾将上肢悬吊，固定于胸前，也可用一条三角巾将伤臂与胸廓环行缚绑住，在对侧胸部打结，再取另一条三角巾将前臂悬挂在胸前。

❷ 前臂骨折

将肘关节屈曲90度，拇指向上，夹板置于前臂外侧，长度超过肘关节至腕关节的长度，然后用绷带将两端固定，再用三角巾将前臂悬吊于胸前呈功能位。

❸ 脊柱骨折

将伤者俯卧于硬板上，避免移动。必要时，用绷带将其固定于木板上。

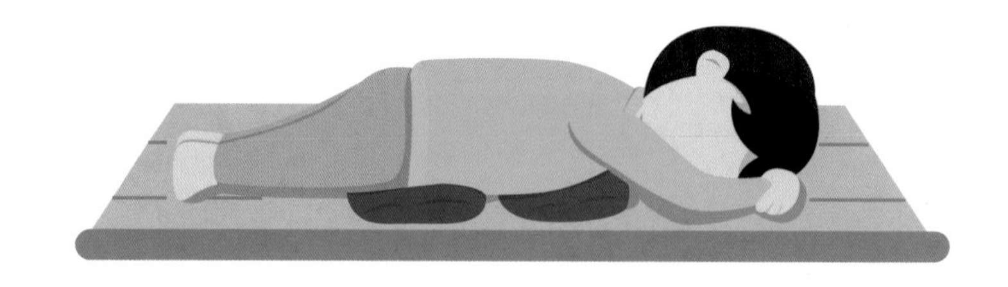

❹ 骨盆骨折

伤者仰卧位，膝微曲，在两膝、两踝之间及下部放一衬垫，后在踝关节、膝关节及髋关节上用三角巾或绷带固定。

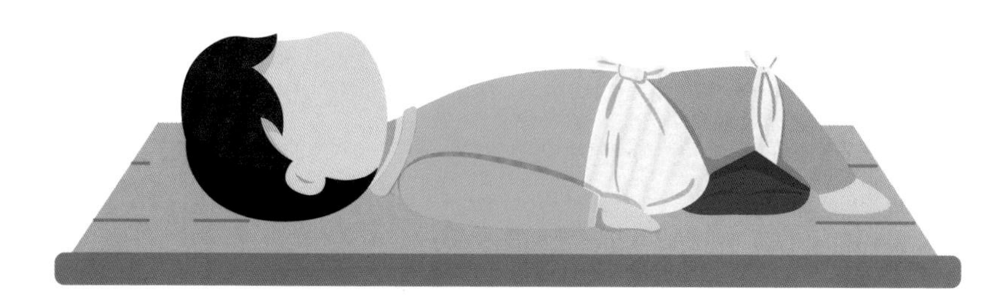

❺ 大腿骨折

取一长夹板置于伤腿的外侧，长度从腰部或腋窝至足跟，另用一夹板置于伤腿的内侧，长度从大腿根部至足跟，然后用绷带或三角巾分段将夹板固定，注意在关节和下肢间的空隙处垫以纱布或其他柔软织物。

❻ 小腿骨折

将两块长短相等的夹板（从足跟至大腿）分别放在伤腿的内外侧，然后用绷带或三角巾分段扎牢。紧急情况下无夹板时，可将两下肢并紧，两脚对齐，然后将健侧肢体与伤肢分段用绷带或三角巾固定在一起，注意在关节和两小腿间的空隙处垫以纱布或其他柔软织物。

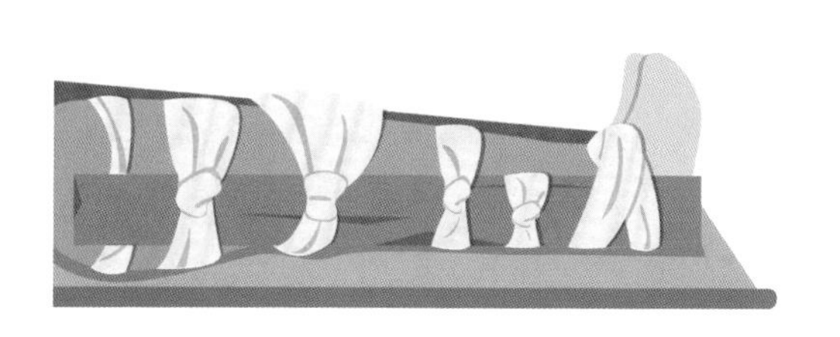

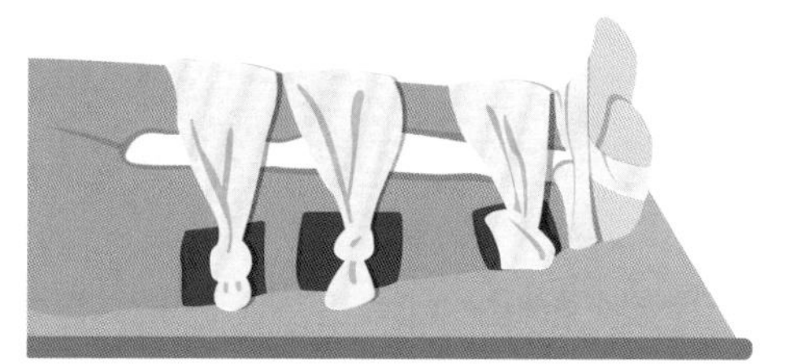

Fixation

When a bone fracture occurs, the injured part should be temporarily fixed to relieve pain and to prevent a secondary injury.

1. Equipment

A splint can be used for a fracture fixation. Strip-shaped hard objects, such as a stick or even another healthy limb can be used in an emergency for fixation.

2. Common methods of fixation

(1) Fracture of the upper arm: Place the splint on the outer side of the injured arm, covering the upper and lower ends of the injured part. Flex the elbow across the body in the most comfortable position. Use a cravat bandage as a swathe to suspend the upper limb and bind it to the chest.
(2) Forearm fracture: Flex the elbow joint to 90 degrees, with the thumb up. Place the splint outside the forearm, the splint should be long enough to reach the elbow joint and the wrist joint. Fix both ends with a bandage, and then use a cravat bandage as a swathe to suspend the forearm and bind it to the chest in a functional position.
(3) Spine fracture: Lie the patient down on a hard board in a prone position, do not move the patient.
(4) Pelvic fracture: Lie the patient down on their back, with their knees slightly bent. Place pads between the knees, between the ankles and under the knees and ankles, then use a cravat bandage or a bandage to fix the position of the ankle joints, knee joints and hip joints.

(5) Fracture of the thigh: Place a long splint along the outer side of the injured leg (long enough to reach from the waist or the armpit to the heel), place a short splint along the inner side of the injured leg (long enough to reach from the end of the thigh to the heel), then fix these two splints with bandages or cravat bandages (Note: place pads of gauze or other soft fabric in the space between the joints and the lower limbs).

(6) Fracture of the calf: Place two splints with equal length (long enough to reach from the heel to the thigh) along the outer and inner sides of the injured leg separately, and fix with bandages or cravat bandages segmentally.

If there is no splint available in an emergency, bring the uninjured leg alongside the injured one; keep the feet aligned. Insert padding between the lower legs. Tie a figure-of-eight bandage around the feet and ankles. Tie the healthy limb and injured limb together with segmental fixation using bandages or cravat bandages, with the knot on the uninjured side.

柳浪闻莺

CHAPTER

05 / 搬运

Transferring Patients

搬运是急救医疗不可或缺的重要组成部分。正确、稳妥、迅速地将伤者搬运至安全地带并对伤者进行抢救、治疗，对预后都至关重要。

5.1 用物

徒手搬运不需任何工具。搬运最常用的器械为担架。现场急救也可用椅子、门板、毯子等代替担架。

注意事项

- 评估伤者的伤势、体重、需要运送的路程、施救者的体力以及其他可能遇到的困难，再进一步选择适当的搬运方法。
- 搬运过程中注意保持平衡，避免操之过急。
- 尽可能动员他人协助，搬运之前务必让所有参与人员明确搬运步骤。

5.2 搬运方法

5.2.1 徒手搬运

救护人员只运用技巧徒手搬运伤者，包括单人搀扶、背驮、双人搭椅、拉车式及三人搬运等，适用于病情轻、路途近又找不到担架的情况。

1 单人搬运法

扶持式 适用于：意识清醒，上肢没受伤能站立行走的伤者。

（1）救助者站在伤者一边，使其手臂揽着自己的头颈，然后用外侧的手紧握伤者的手或手腕。

（2）另一只手伸过伤者背部搂住他的腰部，让他的身体略靠着自己，扶着行走。

抱持式 适用于：年幼伤者、体重轻没有骨折的伤者。

（1）救助者站在伤者一边，一手托其背部，一手托其大腿后抱起。

（2）伤者若清醒，可让他用手抱住救助者的颈部。

背负式 适用于：老幼、体重轻、意识清醒的伤者。

不适用于：胸部创伤、上下肢及脊柱骨折者。

（1）救助者蹲在伤者正前方，微弯背部，让伤者将双臂伸到救助者胸前，将伤者背起。

（2）如伤者卧于地上，不能站立，则救护者可躺在患者一侧，一手紧握伤者的手，另一手抱其腿，用力翻身，使其背负于救护者背上，而后慢慢站起。

❷ 双人搬运法

双人扶助式 适用于：意识清醒上肢无受伤、能合作的伤者。

（1）两名急救者分别站在伤者两旁。

（2）伤者两手搭住急救者的肩膀，急救人员紧握他的手或者手腕。

（3）开步时候，注意步调协调一致。

椅托式

双人坐式：适用于意识清醒但软弱无力的伤者。

（1）两名急救人员分别蹲在伤者两旁，各伸出一只手在伤者的背后交叉支持伤者背部，另外一手伸入伤者大腿之下互相紧握。

（2）尽量将身体贴近伤者，背部保持挺直，慢慢站起，然后一起齐步，外脚先行。

双人轿式： 适用于意识清醒上肢没受伤能合作的伤者。

（1）两名急救者分别蹲在伤者两旁，在伤者的腿下各自以右手握紧自己的左腕，然后用左手紧握对方的右腕。

（2）让伤者两手搭住急救者的肩膀。

（3）尽量将身体贴近伤者，背部保持挺直，慢慢站起，然后一起齐步，外脚先行。

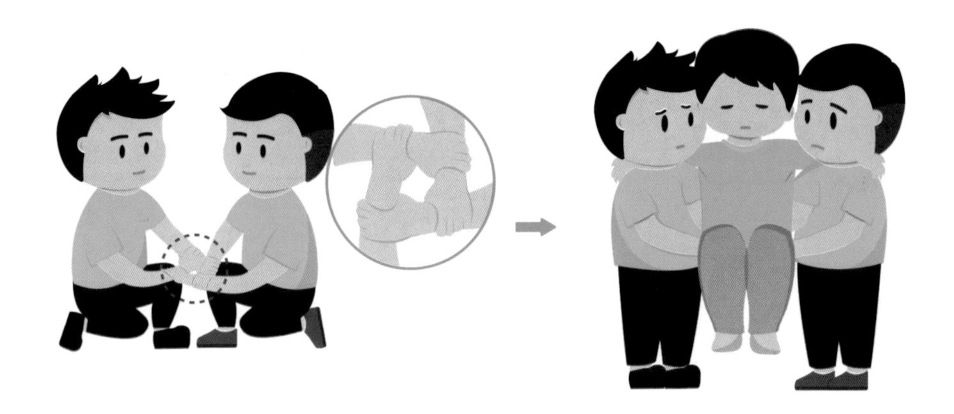

拉车式

双人鞍式：适用于无骨折的伤者（禁用于前臂或者肩部受伤伤员）。

（1）一名救助者在伤者背后蹲下，两手插到伤者腋前，抓紧伤者手腕及前臂，将其抱在怀里，另外一名救助者在伤者两腿中间蹲下，背对伤者挽紧其双腿。

（2）两人慢慢将伤者抬起，步调一致前进。

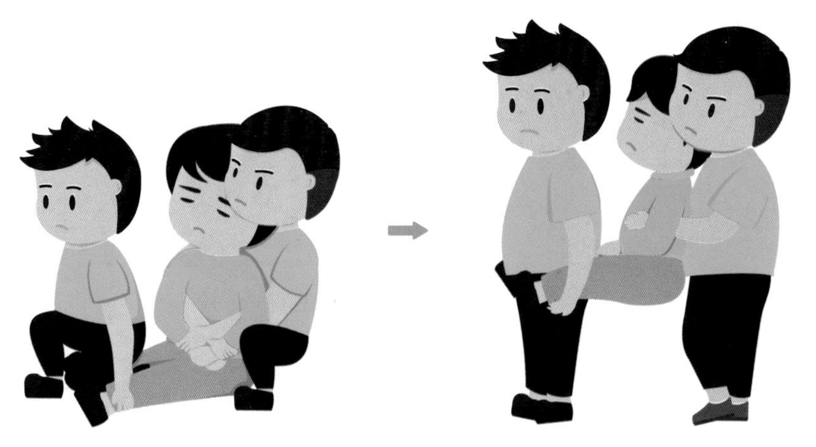

双人平抱式 适用于：胸腰部受伤者。

（1）两名救助者在伤者同侧蹲下，一名救助者将一手插到伤者肩胛背部，一手伸入脖子后方抱住头部，另一名救助者将一手托住腰臀部，一手托住双下肢。

（2）同时将伤者抱起，往怀里靠近。

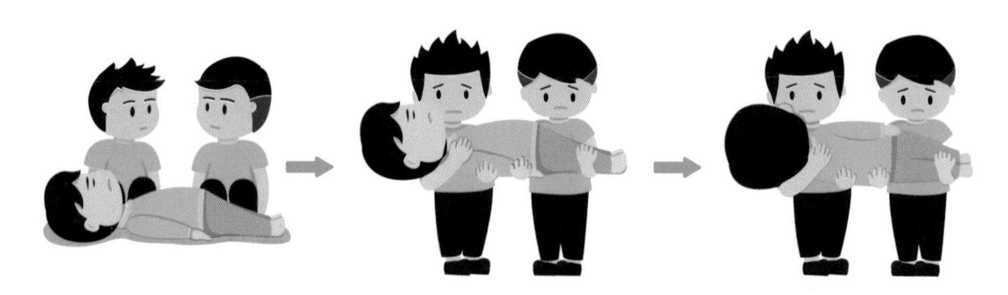

双人对面平抬式

两名救助者分别站在伤者两侧，一左一右站立，一名救助者将一手插到伤者肩胛背部，一手伸入脖子后方抱住头部，另一名救助者将一手托住腰臀部，一手托住双下肢，同时将伤者抱起。

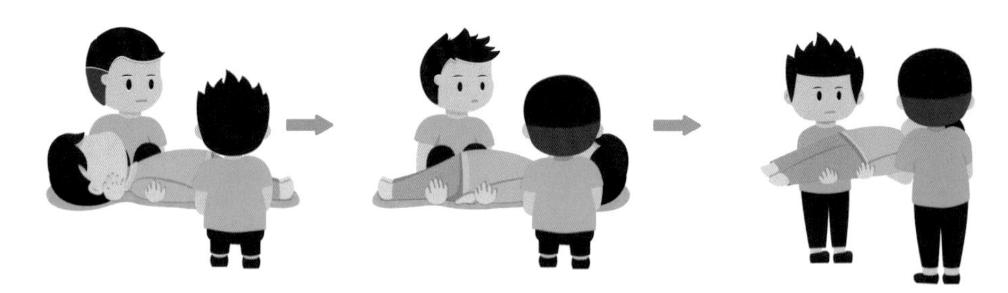

3 三人搬运法

适用于：体重较大者、胸腰部外伤者。

（1）三人平排于伤者一侧蹲下，一人托住肩胛胸部，一人托住腰臀部，一人托住双下肢。

（2）三人同时发力将伤者慢慢抱起，齐步一致前进，注意轻抬轻放。

❹ 多人搬运法

（1）四人或以上救助者分别站在伤者两侧蹲下，一人托住肩胛胸部，一人托住腰臀部，一人托住双下肢，对面救助者方法相同。

（2）大家同时发力将伤者慢慢抱起，面对面站立，齐步一致前进，注意轻抬轻放。

5.2.2 担架搬运

担架搬运最常用，较舒适平稳，一般不受道路、地形限制。

适用于：病情重和运送远途伤者，如有昏迷、内脏损伤、脊柱损伤、骨盆骨折、双下肢骨折等伤势较重的伤者。

可就地取材，用绳索、竹竿、梯子、被服、椅子、门板、木板、毯子等制成简易而结实的担架。

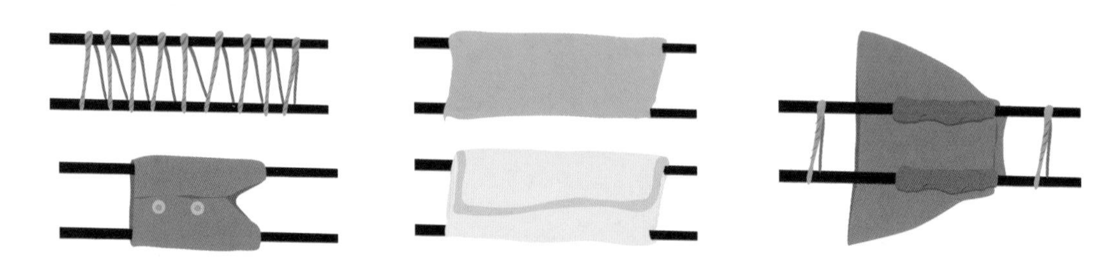

（1）两位救助者站在伤者一侧，一人托住头颈部与腰部，一人托住臀部与双下肢。

（2）将伤者同时水平托起，放在担架上平躺，固定在担架上，注意轻抬轻放，放置时头朝后，脚朝前，方便后面的救助者随时观察伤者的情况。

（3）两人慢慢抬起担架，步调一致前进。

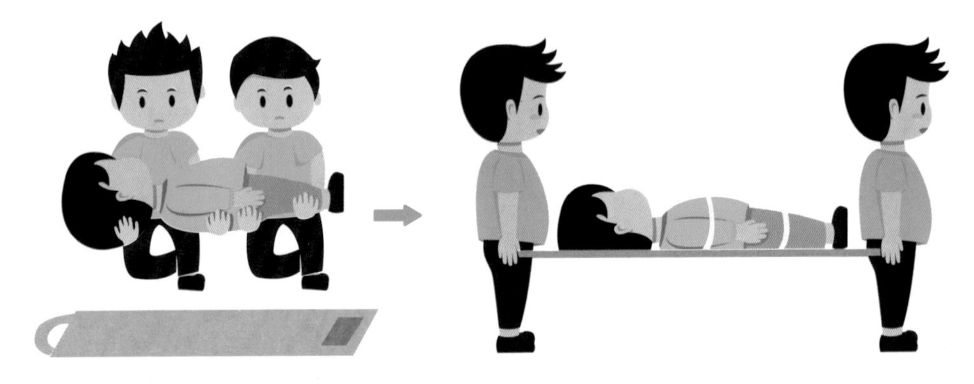

搬运途中尽可能使担架保持水平，使伤者头部保持高位。上坡时，脚放低，头抬高；下坡时则相反。病情若有变化，应立即停下抢救。

5.2.3 特殊伤员的搬运

❶ 脊柱损伤

搬运方法是应先固定颈部，再用硬板搬运。搬运时应有3至4人一起搬动，严防颈部和躯干前屈或扭转，且应保持脊柱伸直，避免加重脊柱、脊髓损伤，保护呼吸功能等。

❷ 骨盆损伤

用大块包扎材料将骨盆作环形包扎，仰卧于硬板或硬质担架上，膝微屈，下面加垫。

❸ 腹部内脏脱出

可用适当的碗或盆等容器扣住伤者内脏脱出部分，并用三角巾包扎固定，屈曲下肢，腹肌放松，并注意腹部保温，严禁将脱出的内脏纳回腹腔，以免引起感染。

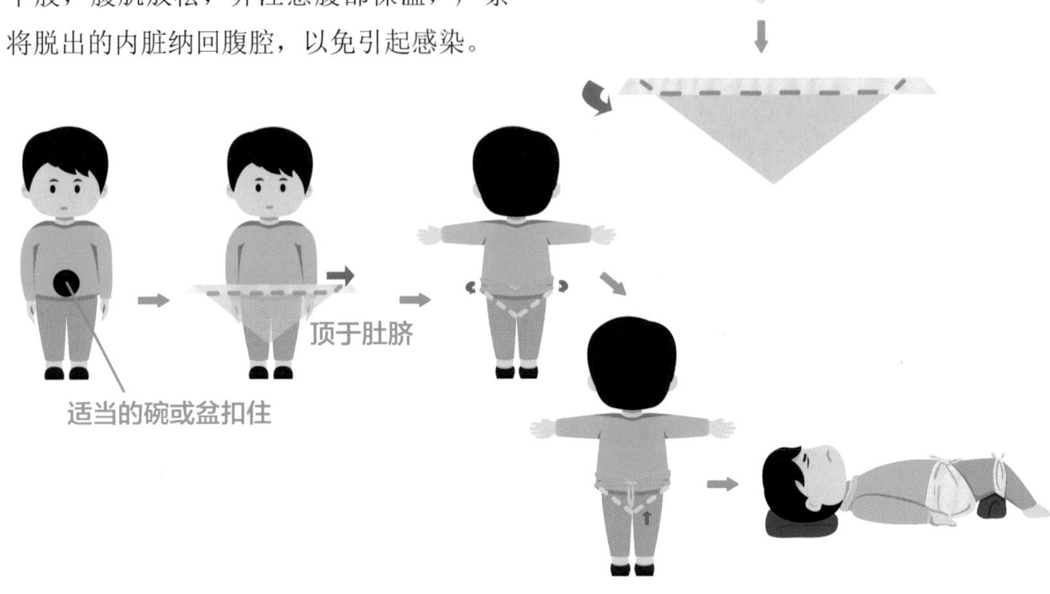

❹ 身体带有刺入物

应先包扎好伤口，固定好刺入物，方可搬运；避免挤压、碰撞；刺入物外露部分较长时，要有专人负责保护刺入物；途中严禁震动，以防止刺入物脱出或深入。

❺ 颅脑损伤、昏迷或有恶心呕吐

使伤者侧卧或俯卧于担架上，头转向一侧，以利于呼吸道分泌物排出。

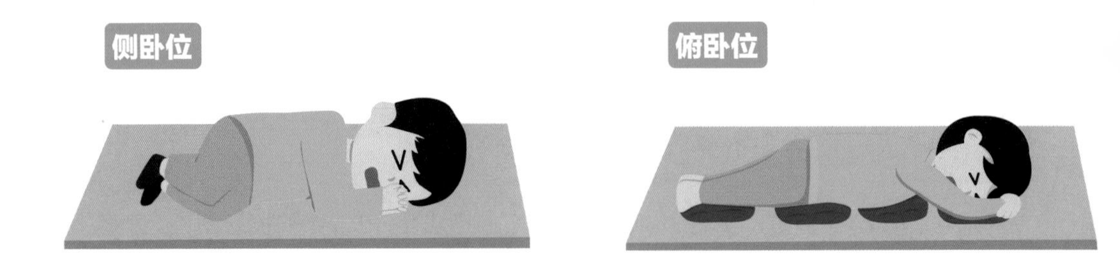

Transferring Patients

Transferring is an important technique used to transport patients to safe areas quickly.

1. Equipment

Transferring patients can use only your hands. Equipment such as stretcher, wheelchair, chair, rope, door panel, can also be used.

2. Methods of transferring patients

Bare-handed transferring is used to transport mildly injured patients to a nearby place when no stretcher is available. It includes lifting the victim by the arms, carrying over the shoulder, two-person seat carry, or a hammock carry, etc.

(1) Transferring by one person.

① The rescuer raise the patient with one arm around the patient’s waist.

It is used to transport conscious patients who have no injuries to any upper limbs, and can stand and walk. The rescuer and the patient stand side by side. The rescue’s another hand grasps the patient's arm which is around his or her head.

② Lift the patient up.

It is used to transport light patients who have no fractures. The rescuer stands beside the patient, and puts one hand on the back of the patient, the other hand under the thighs of the patient, and then lifts the patient up.

③ Carry over the shoulder.

It is used for conscious and not heavy patients, but use cautiously for those with trauma in chest, fracture in limbs and spine.

(2) Transferring by two persons.

① Raise the patient by holding the rescuers' hands or wrists.

It is used to transport conscious patient who have no injuries to any upper limbs, and can cooperate with the rescuers. The patient puts his or her two arms around the rescuers' shoulders, with the rescuers holding the patient's hands or wrists. The two rescuers should walk at the pace.

② Seat carry (four-handed carry).

It is used to transport conscious patient. The two rescuers squat on both sides of the patient, puts one arm across the patient's back, and puts the other under the patient's thighs. Interlock the arms across the patient's back and under the patient's thighs. The rescuers should keep their body close to the patient and their backs straight, stand up slowly, and then walk at the same pace with the outer feet going first.

③ Two-person (or four-handed) cross-bridge carry.

It is used to transport conscious patient who has no injuries to any upper limbs, and can cooperate with the rescuers. The two rescuers squat down at each side of the patient, they grasp their left wrists with their own right hands and grasp the right wrists of each other with their left hands under the patient's thighs, and build a cross-bridge. The patient puts his or her two arms around the rescuers' shoulders.

④ Transfer like pulling a car.

It is used to transport patient without a fracture (contraindicated in those who have an injury in the forearm or shoulder). One rescuer squats down behind the patient, then puts his or her arm across the patient's armpits, grasps the patient's wrists or forearms, and takes the patient into his or her arms. The other rescuer squats down between the two legs of the patient, lifting the patient's legs with his or her back to the patient. The two rescuers walk slowly at the same pace.

⑤ Transfer by cradling the patient with two people.

It is used to transport patient who have an injury to the chest or waist.

- The two rescuers squat down at the same side of the patient, one rescuer puts one of his or her arms under the patient's shoulder, then puts the other arm behind the patient's neck, and holds the patient's head. The other rescuer holds the patient's waist and buttocks using one hand, and holds the patient's lower limbs using the other hand.
- The two rescuers lift the patient at the same time, and bring the patient closer to them.

⑥ Transferring with two people standing face to face.

The two rescuers stand at each side of the patient, one on left and the other on right. One rescuer puts one of his or her arms under the patient's shoulder, then puts the other arm behind the patient's neck, and holds the patient's head. The other rescuer holds the patient's waist and buttocks using one hand, and holds the patient's lower limbs using the other hand. The two rescuers lift the patient at the same time.

(3) Transferring with three people.

It is used to transport heavy patient who has an injury to chest or waist.

- The three rescuers stand on the same side of the patient.The first rescuer holds the patient's shoulders and chest; the second rescuer holds the patient's waist and buttocks; the third rescuer holds the patient's lower limbs.
- The three rescuers lift the patient slowly and walk forward at the same pace. Handle with care.

(4) Transferring with more than three people.

The four or more rescuers stand face to face on each side of the patient, each of them hold the patient's shoulder and chest, or hold the patient's waist and buttocks, or hold the patient's lower limbs. The rescuers standing at the other side have the same assignment.

(5) Transferring using the stretcher.

It is used to transport severely injured patient. The rescuers can make simple but sturdy stretchers by using local materials such as rope, bamboo, a door panel, and board.

① The two rescuers stand at the same side of the patient, one rescuer holds the patient's head, neck and waist, the other rescuer holds the patient's buttocks and lower limbs.

② The two rescuers lift the patient in horizontal at the same time, lay the patient flat on the stretcher with head close to the rescuer walking behind, and feet close to the rescuer walking ahead, so that the rescuer behind can observe the patient's condition at any time.

③ When pulling a stretcher uphill, the rescuer close to the patients' head walks ahead, and the rescuer close to the patients' feet close walk behind, raising the feet to keep it horizontal with the head while transferring; when pulling a stretcher downhill, vice versa.

④ If the patient's condition changes, the rescuers should stop transferring immediately to rescue the patient.

(6) Transferring patient with special condition.

① Patient with spinal injury: Fix the neck first, and then transfer using a hard board to prevent flexion and torsion of the neck and trunk.

② Patient with pelvic injury: Place the patient supine on a hard board or hard stretcher after applying circular bandage around the pelvic.

③ Abdominal visceral prolapse: Cover the prolapse part using a bowl, and fix the position with a cravat bandage. Flex the lower limbs to relax the abdominal muscles, and keep the abdomen warm. It is strictly prohibited to put the visceral prolapse back into the abdominal cavity, lest it cause infection.

④ Penetrating trauma: The rescuer should dress the patient's wound first, and secure the impaled object in place, so that the object won't move and cause further injury, and then transport the patient. Extrusion and collision should be avoided; someone should take care of the foreign object if the part outside the body is long. No shaking during transportation to prevent the object from getting off or penetrating deeper.

⑤ Patient with head injury, coma, nausea or vomiting: Place the patient supine or lateral on a stretcher, with head to one side, to help the secretions of the respiratory to discharge.

黄龙吐翠

CHAPTER

06 气道异物

Airway Foreign Bodies

气道异物梗阻是指异物不慎被吸入气管、支气管所产生的一系列呼吸道症状，重者可造成窒息和死亡。学习气道异物处理技术，使我们能够自救及在他人危急时及时给予正确帮助。

6.1 认识气道

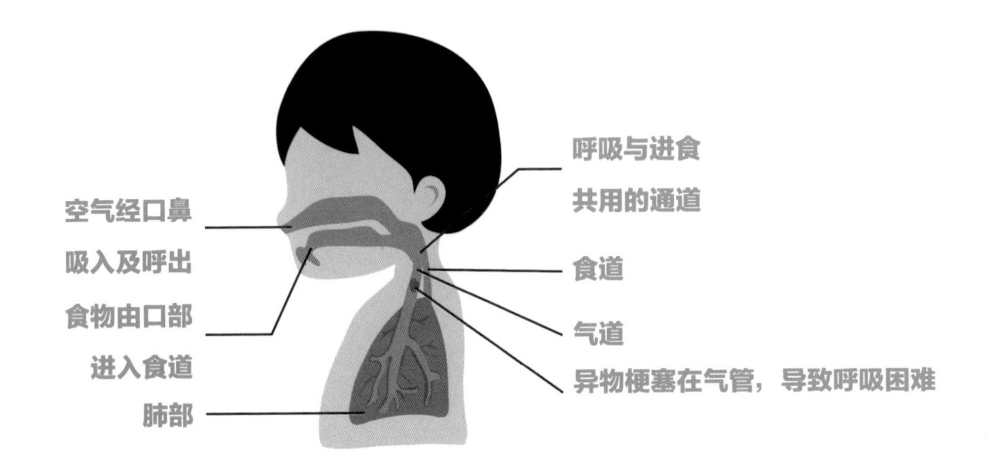

当你看到这样姿势的人，用手“V”字形抓住自己的颈部、喉部，请停下脚步，判断情况后，对其实施急救，能挽救他的生命。

在遇到气道有异物吸入后，首先要判断，千万不能立马去拍背，盲目拍背会使气管里的异物往更深处移动。

6.2 气道异物梗阻喜欢光顾哪些人

吃饭时哈哈大笑

醉酒呕吐误吸

老年人吞咽功能较差

三岁以下宝宝吃东西时
哭笑、打闹等

6.3 救护方法

对气道异物严重梗阻（表现为不能咳嗽、说话或呼吸），在急救的同时，让旁观者拨打“120”急救电话！

6.3.1 成年人和儿童

① 自救法

适用于：意识清楚的人群。

咳嗽法

伤者如能讲话或能咳嗽时，表明气道没被异物完全阻塞，应鼓励尽力呼吸和自行低头咳嗽，重复进行，直到异物咳出。

腹部手拳冲击法

伤者一手握拳，大拇指顶住胸廓肋骨下方与肚脐之间，另一只手抓住该拳头，快速向里向上挤压，重复进行，直到异物吐出。

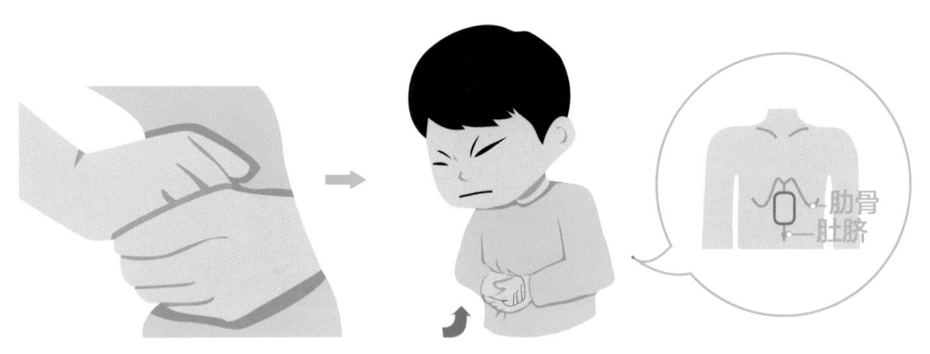

上腹部倾压椅背法

将肚脐以上腹部迅速压在椅背、桌子边缘、扶手栏杆等地，快速向前冲击，重复进行，直到异物吐出。

❷ 手拳腹部冲击法（Heimlich手法）

适用于 意识清楚的人群。

（1）救助者站在伤者背后，让其身体向前倾。

（2）用手臂环绕伤者腰部，一手握拳，大拇指顶住胸骨下方与肚脐之间，另一只手抓住该拳头。

（3）快速向里向上挤压，重复进行，直到异物吐出。

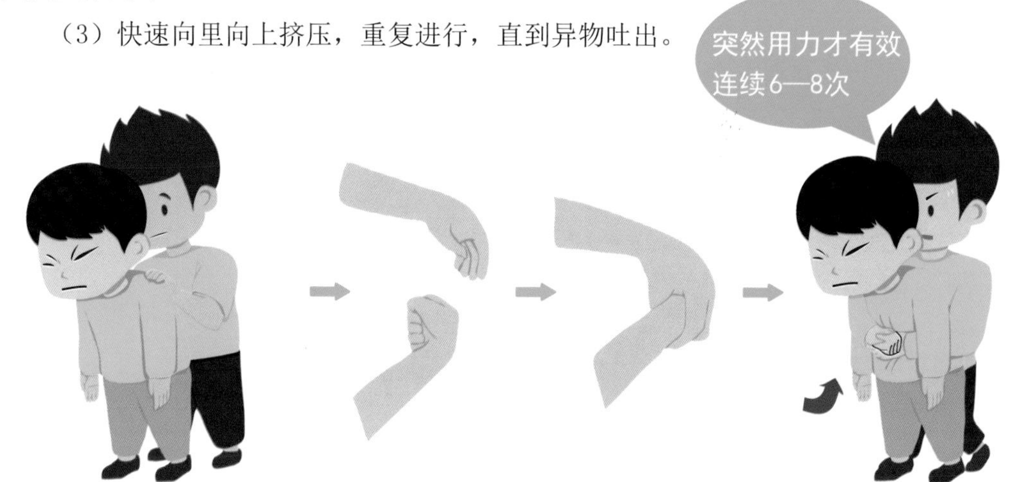

如伤者昏迷，叫不应：

（1）将其平躺在地上，骑跨在伤者髋部，上身前倾。

（2）一手掌根放在伤者胸廓肋骨下方与肚脐之间，另一手放在此手手背上，快速向上向下用力冲击，重复进行，直到异物吐出。

定位要准确，不要把手放在胸骨的剑突下或肋缘下，同时注意胃内容物反流出导致误吸。

③ 手指清除法

适用于 异物卡在咽部以上的昏迷人群。

（1）将伤者按照以下姿势躺好。

（2）救助者一手握住伤者的舌和下颌，另一手食指沿口角内伸入，用勾取动作抠出异物。

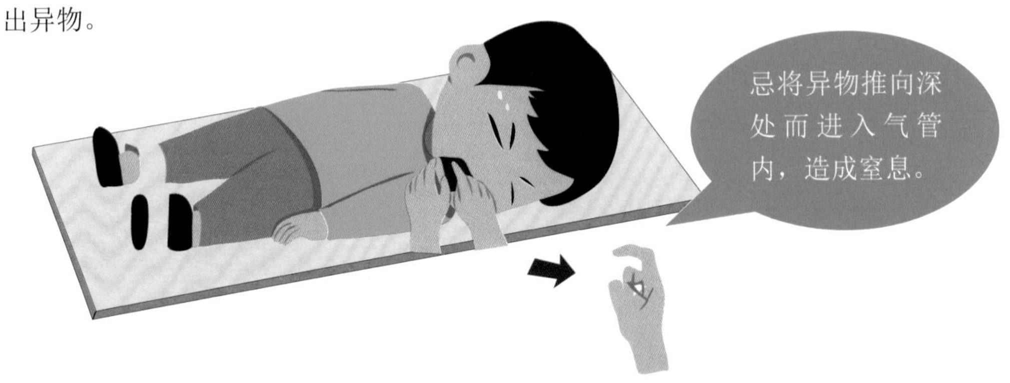

❹ 背部叩击法

（1）将伤者安置成头低背高的姿势，救助者用掌根用力叩击背部（肩胛骨之间）。

（2）头部保持在胸部水平或低于胸部水平。

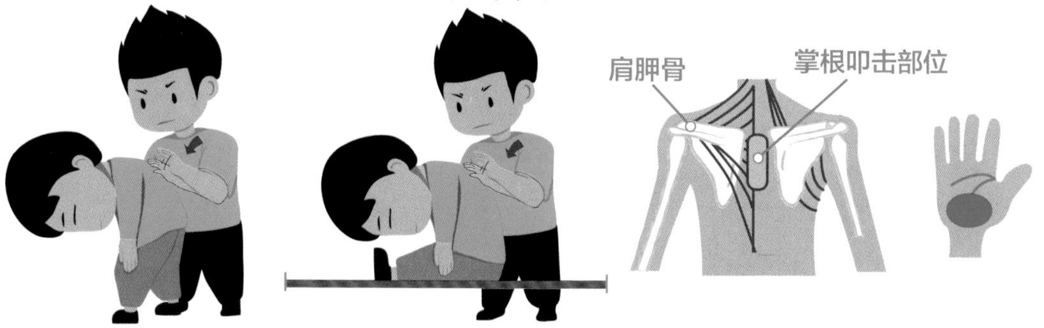

如果仍不能解除梗阻，继续进行6—8次。

如伤者失去意识，应立即转移到坚固的平面上，进行心肺复苏。

6.3.2 婴儿

背部叩击法

第一步

用手扶住宝宝头部，虎口位置放在下颚，胳膊轻放在宝宝身上，另一只手从宝宝脖子后方伸入，手掌固定后头颈背部，慢慢将宝宝翻转方向。

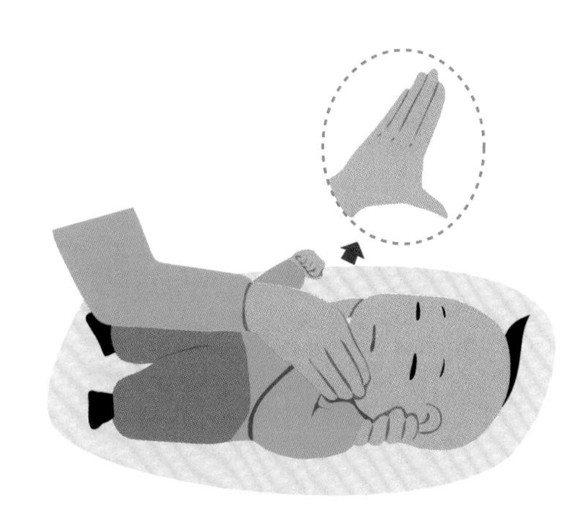

第二步

托住宝宝下颚的手臂放在同侧大腿上，让宝宝趴在手臂上，手掌轻轻托住宝宝下巴，保持头低于身体的位置。

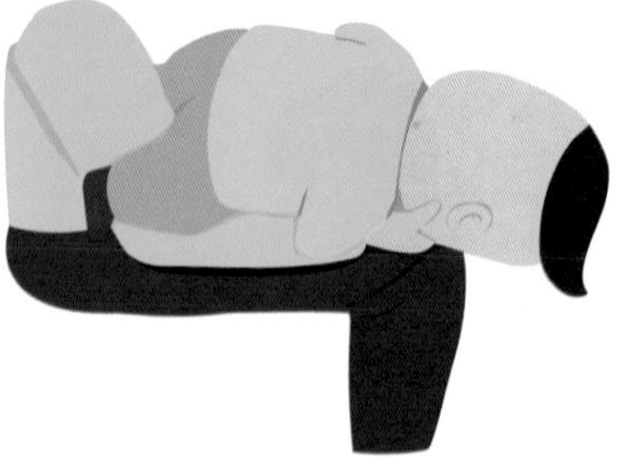

第三步

用托宝宝脖子的手掌根部，连续拍击宝宝的背部（肩胛骨之间），5次后停止。如未能将异物拍出，选择胸部手指叩击法。

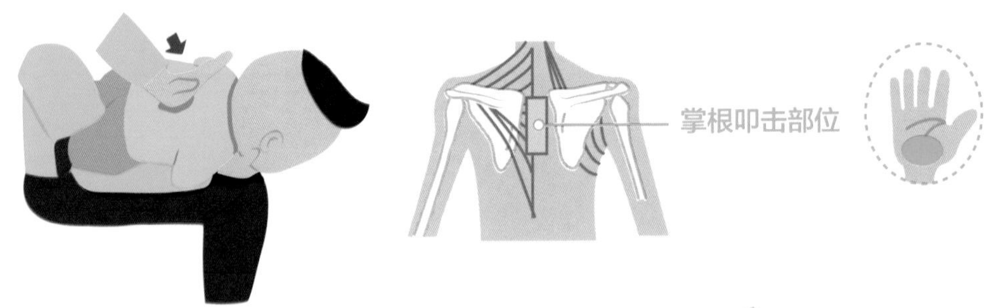

第四步

用拍背的手，将宝宝翻转仰躺在你的手臂上，保持头低于身体的位置。

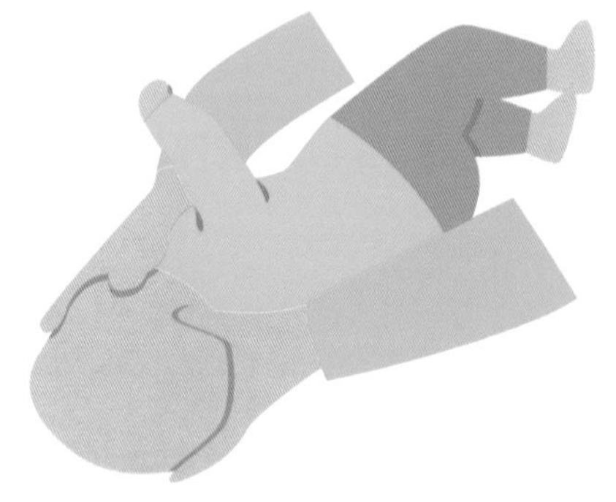

第五步

另一只手的食指和中指对婴儿的胸部（两乳头连线与胸骨中线交界点下一横指处）进行冲击按压，5次后停止。

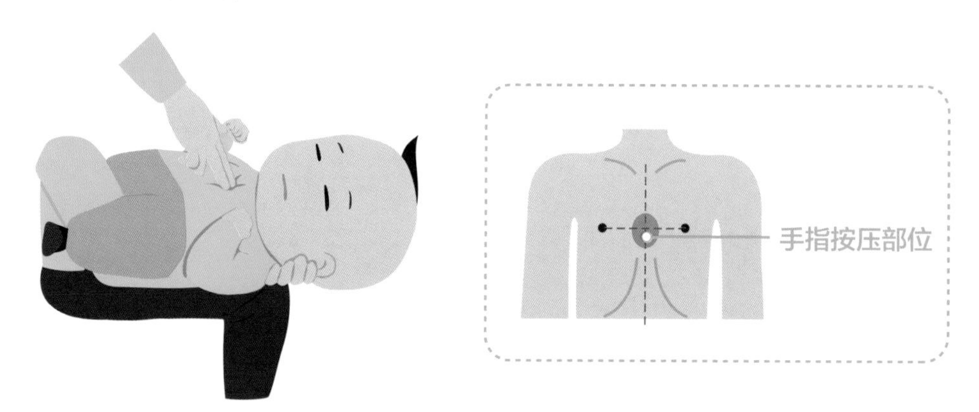

如果仍不能解除梗阻，继续交替进行背部叩击和胸部手指叩击法。

如患儿失去意识，应立即转移到坚固的平面上，进行心肺复苏。

Airway Foreign Bodies

Foreign object airway obstruction means that a foreign body is stuck in the throat, trachea or bronchus. People can't breathe normally which can lead to asphyxia and needs emergency treatment.

1. Sign of choking

If you see a person who desperately clutches his or her neck or throat, please stop to provide first aid.

Don't slap their back immediately, otherwise, the foreign object in the airway may move deeper.

2. Abdominal thrust (The Heimlich maneuver)

Indications: conscious patients.

(1) The rescuer stands behind the patient, and instructs the patient to lean forward.

(2) Place arms around the patient's waist, curl one hand into a fist, and place the thumb side of the fist in the space between the navel and well below the breastbone. The other hand seizes the fist.

(3) Exert quick, forceful upwards thrust on the patient's abdomen. Repeat the action. Repeat thrusts until the object is expelled from the airway or the patient become unresponsive.

3. Choking relief in an unresponsive patient

(1) If the choking patient become unresponsive, shout out for help. Send someone to call 120.

(2) Gently lower the patient to the ground, begin CPR immediately.

(3) Chest compressions. Do not check for a pulse.

(4) Each time you open the airway to give breaths, open the patient's mouth, if you see visible object, remove it; if no, continue CPR.

双峰插云

CHAPTER

07 / 胸痛

Chest Pain

胸痛的原因有很多种，若在体力劳动、情绪激动、寒冷刺激、饱餐后突然出现胸骨后疼痛，持续时间短，疼痛向左肩、颈部、右肩、手臂、背和上腹部、面颊部放射，有压榨感或窒息感，甚至有濒死或恐惧感时，应怀疑为心绞痛、心肌梗死，需引起重视，应尽快到医院进行确诊治疗。

7.1 胸痛

（1）伤者在体力劳动、情绪激动、寒冷刺激、饱餐后突然出现胸骨后疼痛，持续时间短，疼痛向双肩和左臂内侧、左颈部或面颊部放射，伤者有压榨感或窒息感，甚至有濒死或恐惧感，应怀疑为心绞痛，经休息或舌下含服硝酸甘油片缓解。

（2）如上述疼痛持续数小时至数天，伴有口唇、四肢皮肤青紫，则应怀疑为急性心梗。

（3）心前区疼痛剧烈，有紧压感，放射至左肩、左臂内侧、左肩胛区、背部、颈部、下颌部及剑突下，疼痛可呈持续性或间歇性，咳嗽、深吸气、举臂时可使疼痛加剧，多见于急性心包炎。

（4）胸部一侧或双侧烧灼样疼痛，伴有咳嗽、咳痰，体温升高，呼吸时疼痛加剧，多为肺炎；感觉胸部肋间如闪电样一过性疼痛，局部有压痛，则多见于肋间神经痛。

胸痛是一种常见症状，尤其对心血管疾病的诊断有重要意义。由于胸痛的剧烈程度不一定和病情轻重相一致，故发现胸痛，应及时到医院就诊，X线、心电图、超声心动图常能为诊断提供正确的依据。

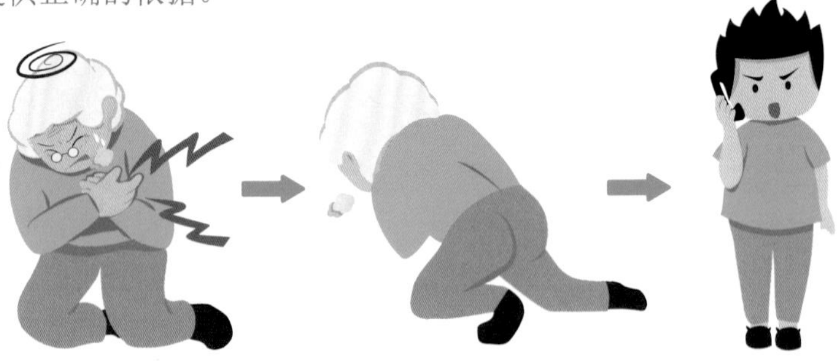

- 做好心理护理，让伤者静卧休息，做好解释工作，消除恐惧，尤其是心绞痛、心肌梗死伤者应绝对卧床休息，以减少心肌耗氧，有利于减轻疼痛。
- 出现胸痛，尤其是老年人，必须考虑有冠心病的可能。若有冠心病史，则诊断更易明确。应嘱其停止活动，休息，舌下含服硝酸甘油等，并尽快联系医生，尽早处理。

Chest Pain

Chest pain may be a symptom of a number of diseases. Pay special attention if the pain is experienced as tightness around the chest while radiating to the left arm and the left angle of the jaw, and pressure-like in character, typically occurring after physical work, being emotional, cold simulation or after a heavy dinner. Sometimes it may be associated with sweating, nausea and vomiting, as well as shortness of breath. Heart attack may occur, and an immediate medical treatment should be considered. While waiting for the arrival of healthcare providers, it is important for the patients to lie down and rest, keep calm and get distracted. Have nitroglycerin under the tongue if possible.

花港观鱼

CHAPTER

08 急性中毒

Acute Poisoning

急性中毒是指大量毒物短时间内经皮肤、黏膜、呼吸道、消化道等途径进入人体，致使机体受损并发生功能障碍。急性中毒是生活中较常见的急症，一般起病急骤，症状严重，病情变化迅速，不及时治疗常危及生命。

8.1 毒物

日常生活中，民众了解中毒的常见症状，掌握基本的急救方法，是一项重要的生活技能。

8.2 中毒途径有哪些

❶ 口服（消化道）

吃了一些带毒食物、药物或其他有毒物质。

❷ 吸入（呼吸道）

吸入有毒气体如煤气、工业废气、浓烟、有机溶剂等。

❸ 皮肤吸收

农药、水银、杀虫剂等。

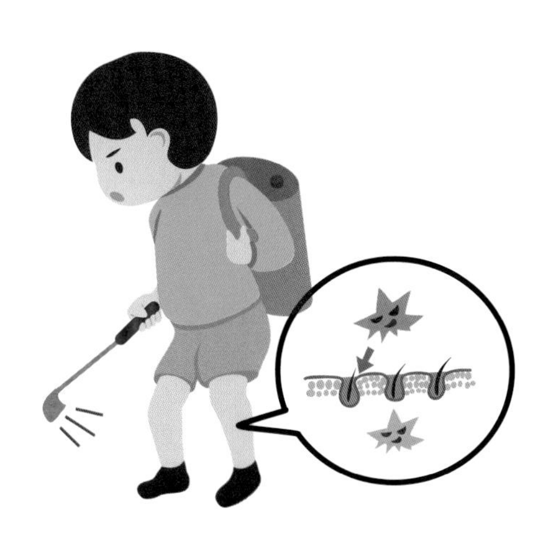

❹ 其他途径

蚊虫叮咬、蜇伤，有毒物通过针管注射到体内等。

8.3 急性中毒的急救原则

- 救助他人前先确保自身安全，做好防护，以防中毒。
- 如伤者已意识不清，立即进行心肺复苏，人工呼吸时注意防护。
- 立即终止伤者接触毒物，清除未吸收的毒物（搬离受污染的地方，比如煤气泄漏现场；脱去或剪去衣物，用大量清水冲洗等）。
- 拨打“120”呼救，说清楚以下几点：伤者是否清醒、清醒程度、有无呕吐、初步提供引起中毒的原因。
- 促进已吸收毒物的排出。
- 尽量保留现场遗留的毒物、药袋及呕吐物等，交给医护人员，以便更快治疗。

8.4 急性食物中毒

8.4.1 引起急性食物中毒的食品

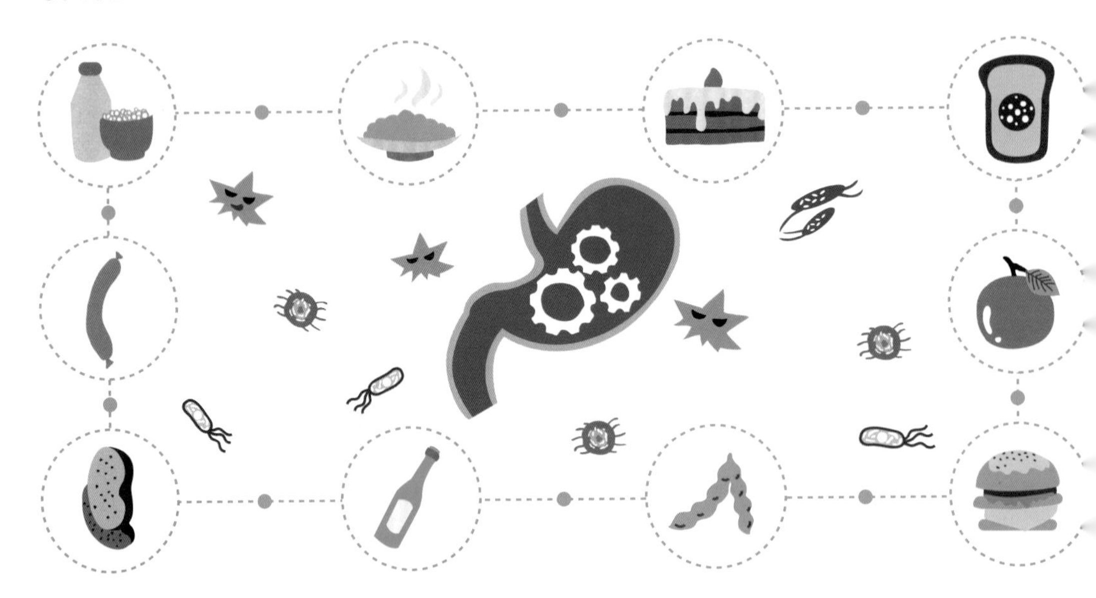

发芽的马铃薯、毒蘑菇、毒鱼、苦杏仁、白果、没煮熟的扁豆、隔夜发酵的食物等。这些食物中有被细菌及其毒素污染的，食物本身具有自然毒性的，也有生吃或没煮熟细菌未被杀死的，还包括未做好保存及清洁工作导致食品污染的。

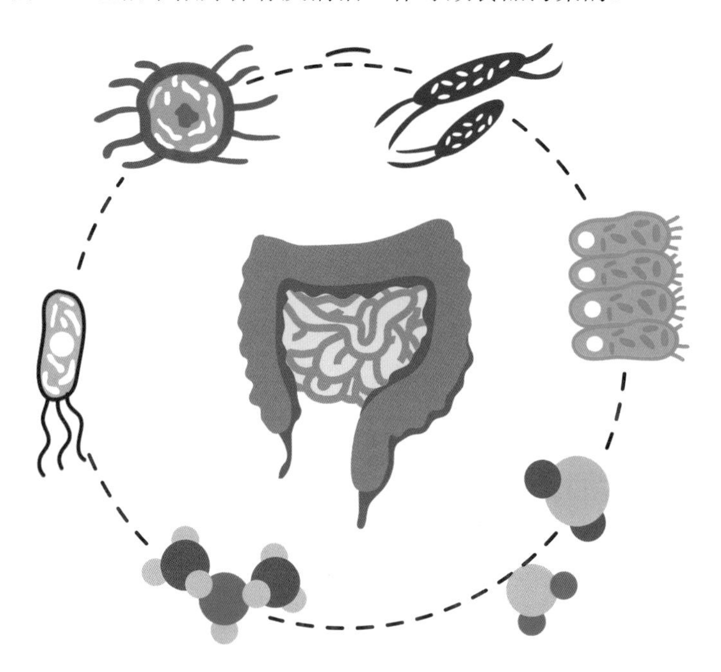

8.4.2 急性食物中毒有哪些表现

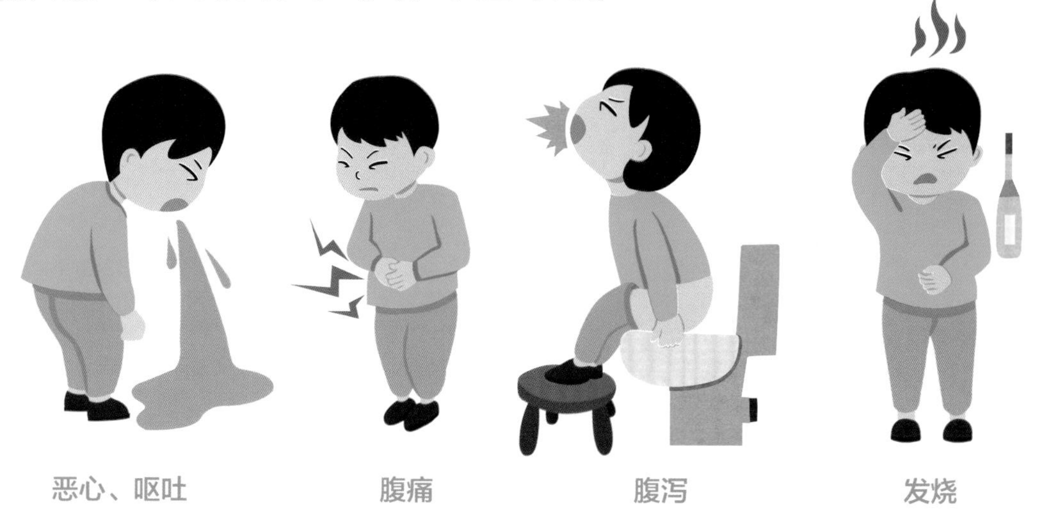

恶心、呕吐　　腹痛　　腹泻　　发烧

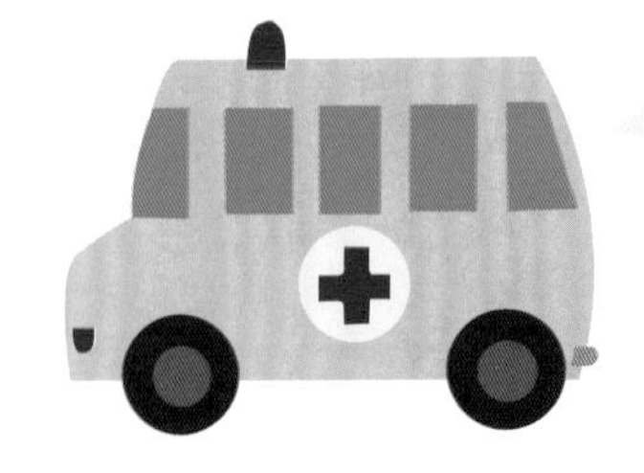

大多发生在夏秋季节，常在食用后1小时至1天内出现：恶心、呕吐、腹痛、腹泻、发烧、头痛，严重者有生命危险。

头痛　　严重者有生命危险

8.4.3 急性食物中毒救护方法

- 意识清醒者大量喝水，并将手指伸入喉咙深处，使胃内食物呕吐出来。
- 口服泻药。
- 尽可能保存伤员的食物样本、呕吐物、排泄物等。
- 将伤者立即送往医院急救。

伸入手指

呕吐

服用泻药

8.4.4 如何预防食物中毒

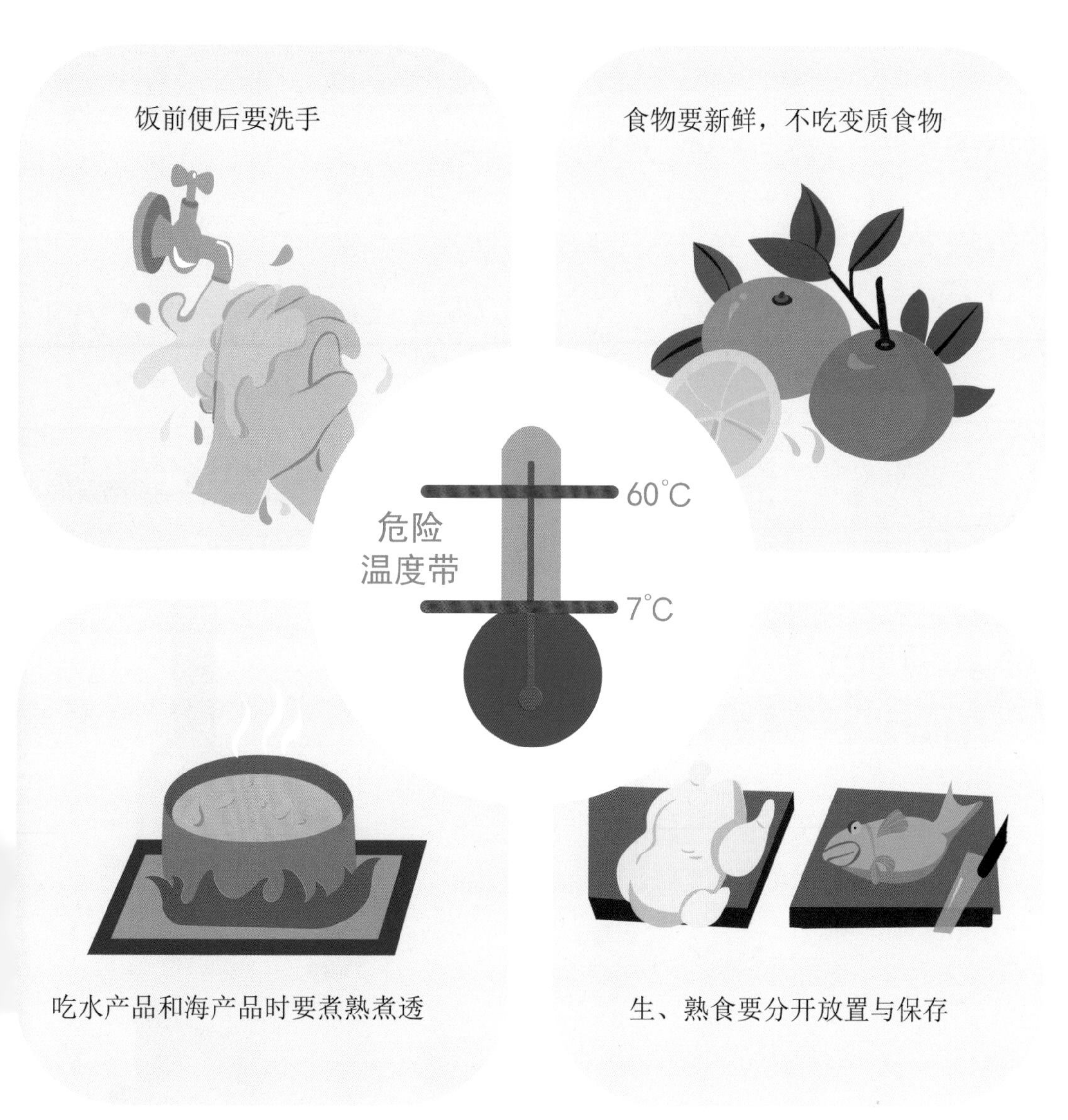

生活小招 吃凉拌菜加醋和大蒜，加醋5分钟就可消灭95%的沙门氏菌！

8.5 酒精中毒

饮酒，是日常生活中常见的消遣方式，是餐桌文化的必备品，但饮酒过量，就很可能引起酒精中毒，对人体造成重大损害。

8.5.1 认识酒精

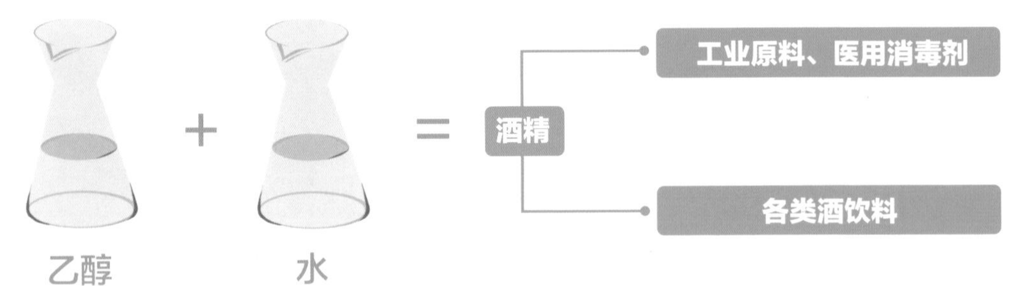

8.5.2 酒精中毒主要原因

大量饮酒或食用高浓度酒精。

酒精的中毒量因人而异，一般为70—80克，若在250—500克则会严重危害生命。

发生中毒是否与下述因素有关：

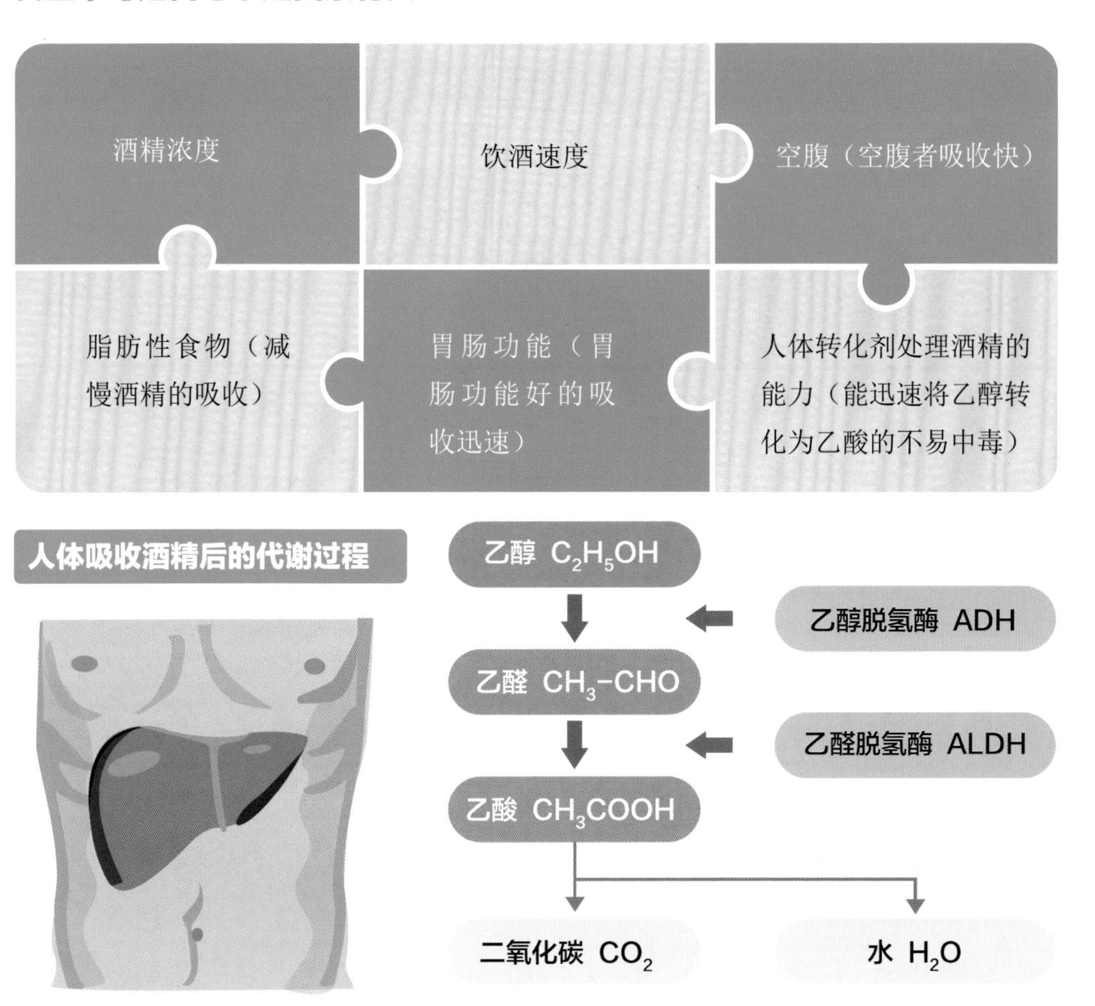

饮酒后的酒精约20%在胃内吸收，约80%在十二指肠及小肠吸收。

乙醛可刺激肾上腺素、去甲肾上腺素等的分泌，表现为面色潮红、心跳加快等。酒精具有直接的神经系统毒性、心脏毒性和肝脏毒性，甚至会引起昏迷及休克，此外还可发生低血糖和代谢性酸中毒。

8.5.3 酒精中毒危害大

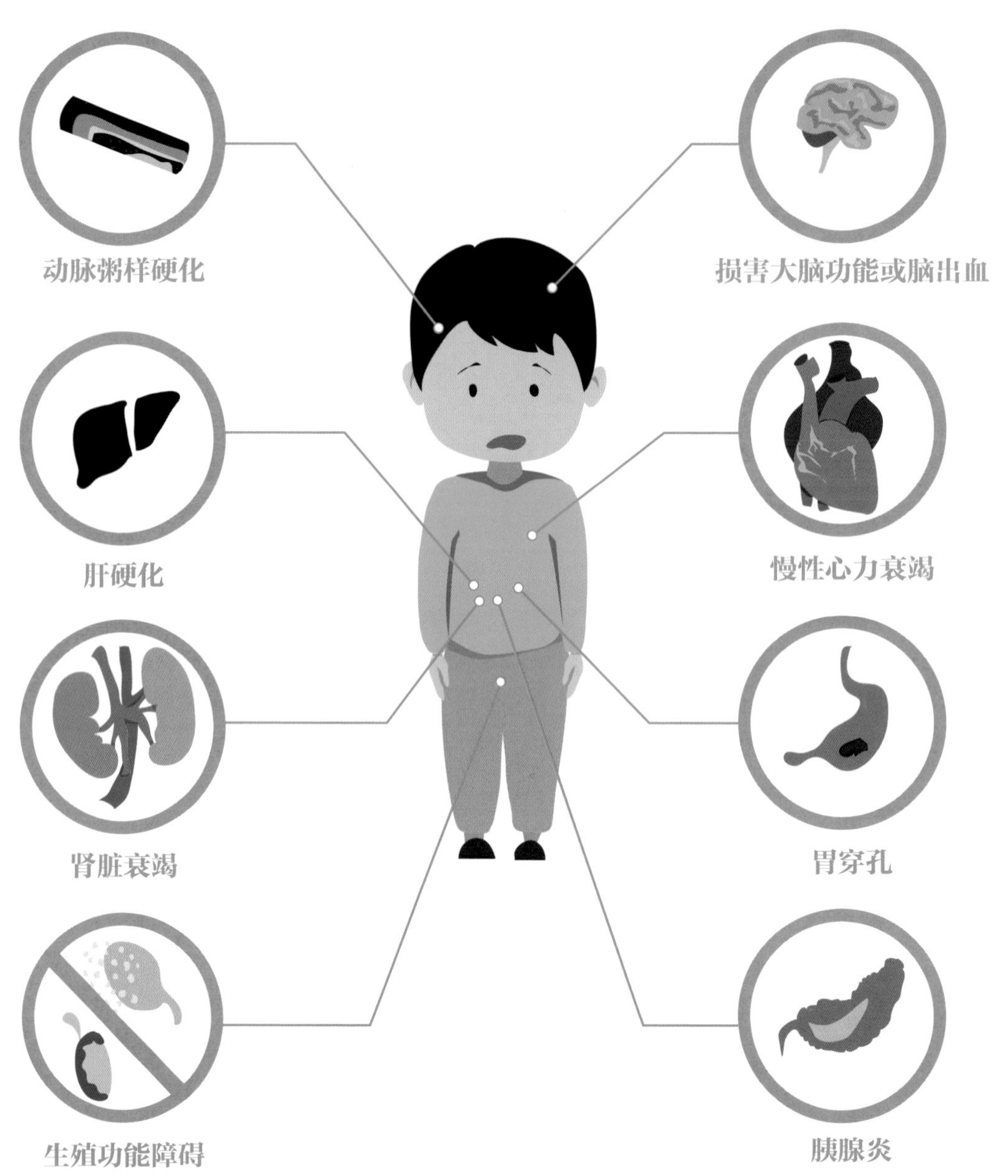

8.5.4 救护方法

（1）轻度醉酒者，可让其静卧，最好是侧卧，以防吸入性肺炎，注意保暖，避免受凉。

（2）重度酒精中毒但神志清醒者，应用筷子或勺把压其舌根部，迅速催吐。

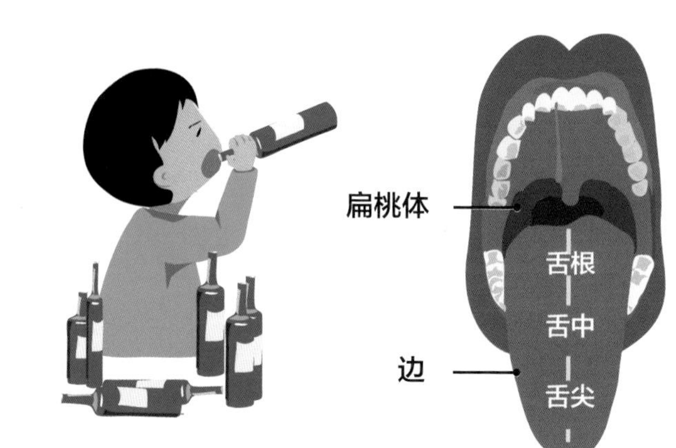

（3）若酒精中毒者昏迷不醒，应立即送医院救治。

8.6 一氧化碳中毒

急性一氧化碳中毒又称为煤气中毒，是人体短时间内吸入过量一氧化碳而造成大脑及全身组织缺氧性疾病，严重者危及生命。发生在家庭中主要是指一氧化碳、液化石油气、管道煤气、天然气中毒等，在冬天多见于使用煤炉取暖、液化灶具泄漏、煤气管道泄漏等情况。

8.6.1 认识一氧化碳

一氧化碳是一种无色无味的气体。一切含碳物质，在燃烧不完全时，都能产生一氧化碳。正因为一氧化碳的存在是无形的，人往往很难警觉，所以更要引起重视。

一氧化碳进入人体血液后与血红蛋白结合形成碳氧血红蛋白，血红蛋白丧失携带氧气的能力，也就不能及时供给全身组织器官充分的氧气，其中影响最为严重的是大脑。

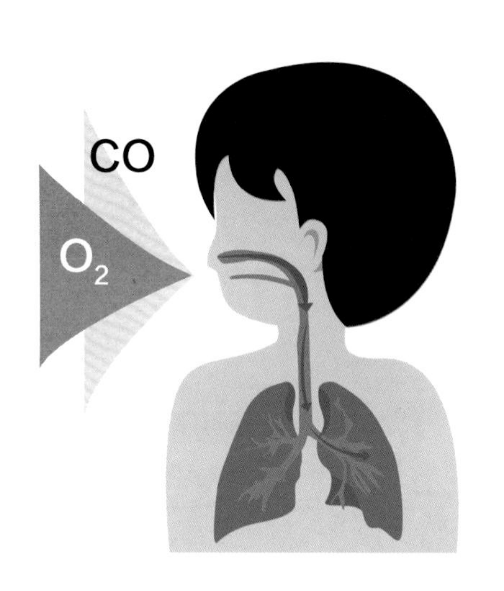

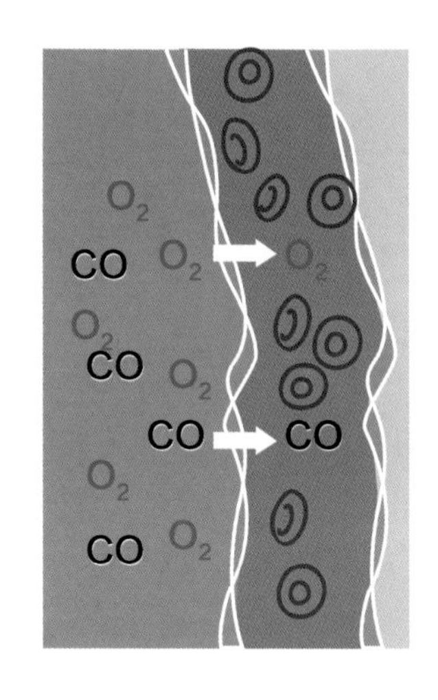

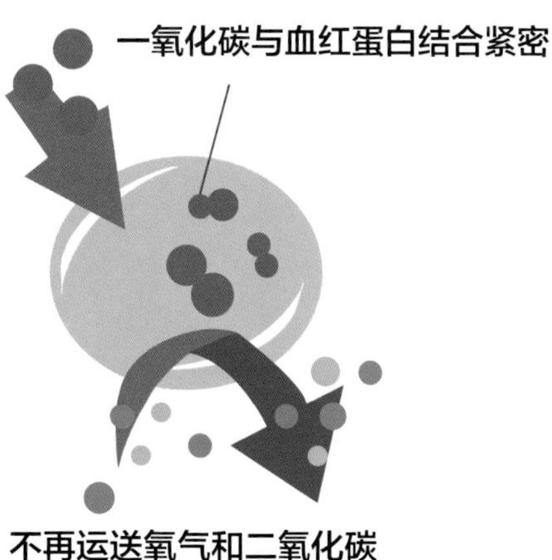

大脑是最需要氧气的器官之一，由于体内的氧气只够消耗10分钟，一旦断绝氧气供应，将很快造成人的昏迷，如不能及时发现并抢救就会危及生命。

8.6.2 一氧化碳中毒程度与表现

中毒程度	血液COHb浓度	表现
轻度中毒	10%~20%	出现轻微头痛、头晕、四肢无力、恶心、呕吐、心悸及视力模糊，如及时脱离中毒环境，吸入新鲜空气后症状迅速消失。
中度中毒	30%~40%	除上述症状外，中毒者嘴唇、指甲呈樱桃红色，呼吸急促，判断力弱、视线模糊、昏昏欲睡等，经治疗可恢复且无明显并发症。
重度中毒	≥50%	意识不清，昏迷，不脱离原先环境将有生命危险。

与空气中一氧化碳、血液中COHb浓度呈正比例关系，也与个体健康情况有关，如怀孕、嗜酒、营养不良、贫血、有心血管疾病和呼吸系统疾病等均会加重中毒的程度。

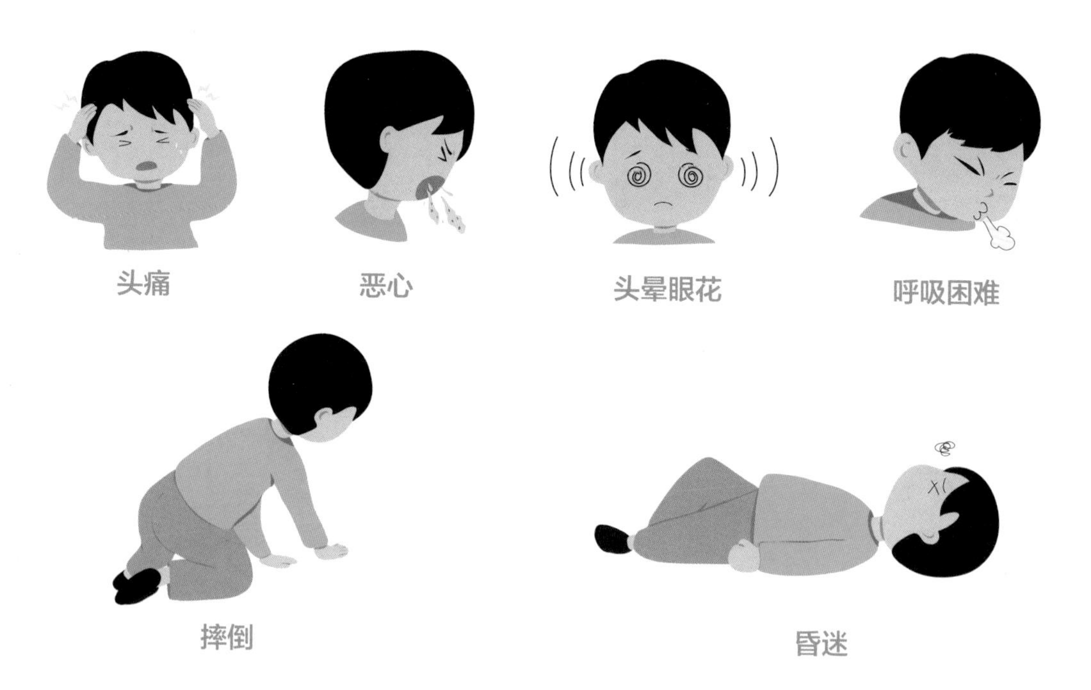

8.6.3 一氧化碳中毒后如何急救

自救

人身处使用煤炉、炭盆或煤气热水器的环境中，出现头晕、眼花、恶心、胸闷等症状时，应关闭煤炉开关，切断中毒源，尽快打开门窗并离开有毒环境。

若感到全身无力不能站立时，要在地上匍匐爬行，迅速打开门逃生并呼救。

救助他人

（1）一旦发现或怀疑有人一氧化碳中毒，用湿毛巾捂住口鼻，匍匐入室内。

（2）迅速将中毒者搬至空气新鲜流通处，解开其衣领和腰带，保持呼吸道通畅。

（3）关闭煤气开关，打开门窗通风。

（4）切记不可开关电器，不使用明火。

（5）立即将中毒者送往医院，如发现呼吸心跳停止，应立即进行心肺复苏，同时拨打“120”急救电话。

8.6.4 如何预防一氧化碳中毒

（1）使用煤气热水器洗澡时，应将液化气罐及热水器放在室外通风处。已在室内的，洗澡时最好不要紧闭门窗。

（2）使用煤炉、炭火烧东西或取暖时，应在通风处，并在睡前将炭火炉移至室外。在使用炭火、液化气等涮火锅的无窗包厢内用餐时，要开启换气扇。

（3）使用煤气时，最好人不远离，以免发生意外。

（4）定期检查连接煤气的橡胶皮管有无松脱、老化、破裂、虫咬等情况，及时更换。

（5）有条件者，尽可能在能产生一氧化碳的地方安装一氧化碳报警器，以便在气体浓度超标时进行有效应对。

（6）开车时，不要让发动机长时间空转；车在停驶时，不要过久地使用车上空调机；即使是在行驶中，也应经常打开车窗，让车内外空气产生对流。感觉不适即停车休息；驾驶或乘坐空调车如感到头晕、发沉、四肢无力时，应及时开窗呼吸新鲜空气。

Acute Poisoning

In daily life, poisons can be swallowed, inhaled, absorbed through the skin, splashed into the eyes, or injected. The symptoms vary from person to person. Whatever the poison is, first aid should be done immediately after the poisoning.

First aid principles for the acute poisoning:

(1) Ensure your own safety first. Take protective measures to prevent poisoning.
(2) If the patient is unconsciousness, CPR should be performed immediately and barrier device should be used when delivering breath.
(3) Terminate contact with the poison immediately, and clear away the unabsorbed poison (remove the patient from the vicinity of the contaminated areas, such as a leaking gas field; remove or cut off the clothing and wash the skin with a lot of water, etc.).
(4) Call 120 for help and clarify the following: whether the patient is awake and the degree of soberness; whether the patient is vomiting; the possible causes of poisoning.
(5) Try to excrete absorbed poison.
(6) Try to keep the poison, medicine bags and samples of vomited material left on the spot, and then give them to the medical staff.

1. Acute food poisoning

Acute food poisoning mostly occurs in summer and autumn, and often appears from one hour to one day after consumption. Common symptoms include nausea, vomiting, an upset stomach, diarrhea, a fever, a headache, and even death in severe cases.

Rescue methods:

(1) If conscious, the patient should drink lots of water, and stretch a finger into the deep throat to vomit food out.

(2) Drink laxatives.

(3) If possible, keep food samples, vomited stuff and excrement of the patient.

(4) Send the patient to the hospital as soon as possible.

2. Alcohol poisoning

Excessive intake of alcohol may cause alcohol poisoning. Cautions should be taken to prevent further risk, such as aspiration, hypothermia, head injury, stroke, heart attack or hypoglycemia.

Rescue methods:

(1) A mildly intoxicated person should have a rest, and it is the best to lie on one's side to prevent aspiration pneumonia. Cover the person with a coat or blanket to keep warm and avoid catching a cold.

(2) People with severe alcoholism but is conscious, use a chopstick or a spoon to press the root of the tongue, letting the patient vomit out the alcohol in the stomach.

(3) If the patient sinks into deep unconsciousness, call 120 for help right away.

3. Carbon Monoxide Poisoning

Acute carbon monoxide poisoning, also known as gas poisoning, is the oxygen deficiency disease in the brain and all over the body, caused by inhaling excessive amounts of carbon monoxide within a short period of time; it can be life-threatening.

(1) Self-help.

If you have symptoms such as dizziness, vertigo, nausea, or chest tightness when in the environment of a coal stove, charcoal basin or gas water heater, you should open windows and doors as soon as possible and leave the toxic environment. If you feel that your whole body is weak and can't stand, you can crawl on the ground, open the door to escape and call for help quickly.

(2) Help others.

Once you find or suspect that someone has carbon monoxide poisoning, you can cover your mouth and nose with a wet towel and crawl indoor. Move the patient to a place with fresh air, and then loosen their collar and belt to keep the respiratory tract unobstructed. Switch off the gas then open windows and doors to ventilate. Don't turn on electrical appliances or use open flames.

The patient should be sent to the hospital as soon as possible. If the patient turns to unconsciousness, or pulse could not be detected, or without normal breathing, you should immediately do CPR and call 120 at the same time.

云栖竹径

CHAPTER

09 / 毒蛇咬伤

Snake Bites

毒蛇咬伤是指毒蛇的毒液经排毒导管进入人体，并经淋巴和血液循环扩散，引起局部和全身中毒的症状，一般常发生在农民、渔民、野外作业者和从事毒蛇养殖业者及研究人员的身上。在战争和野外训练时，战地官兵偶尔也会被毒蛇咬伤。毒蛇咬伤以夏、秋两季为多见，咬伤部位以手、臂、足和下肢为常见。

9.1 如何区分有毒蛇或无毒蛇

区别点	毒蛇	无毒蛇
头部	多呈三角形	一般椭圆形
		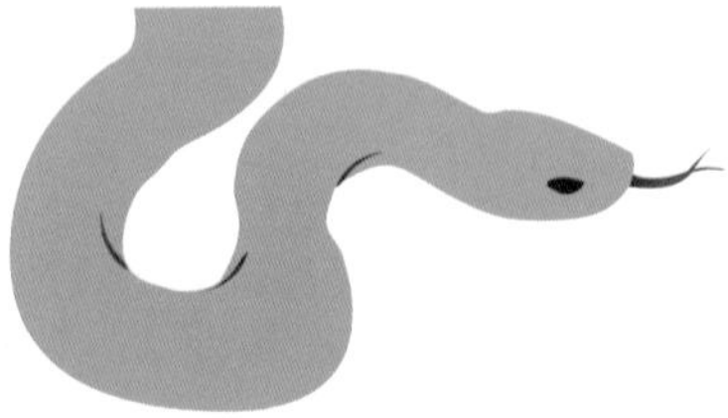
尾巴	短钝或呈侧扁形	长而尖细
体色	颜色较为鲜艳，或有特殊斑纹	多不鲜艳
体型	短而粗，不均匀	长细，体型匀称
毒牙	有毒牙、毒腺	无毒牙、毒腺
	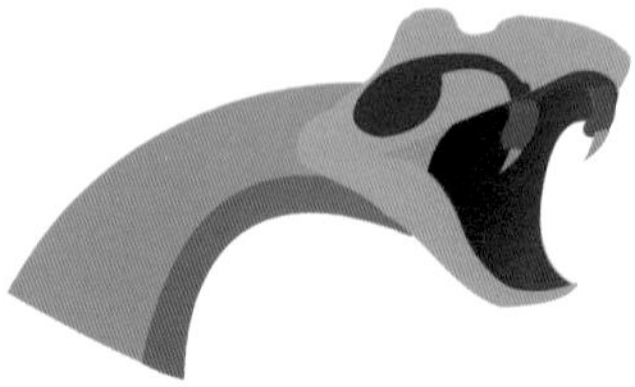	
动态	休息时常盘成团	爬行时动作迅速
性情	凶猛，具有攻击性	胆小怕人，容易受惊
蛇咬伤牙痕辨别		

9.2 在生活中遇到毒蛇咬伤该怎样进行急救

毒蛇毒腺内储存的毒素进入血液循环后数小时，甚至几分钟就可以蔓延全身，引发症状，因此时间是救治的重要因素。

❶ 保持镇静

惊慌失措不仅对救治没有丝毫益处还会延误时机，此时伤者应保持冷静，并且不能奔跑，因为运动会使血液循环加速从而促进毒素扩散。同时求助120，联系最近有抗蛇毒血清的医院。

❷ 结扎

可以就地取材，利用鞋带、布条、手帕或是路边的稻草、树皮在伤口上方1—5厘米、靠近心脏的一端进行结扎，松紧度以日常抽血化验时橡胶管绑的程度为宜。每隔15—20分钟需放松1—2分钟，防止肢体因为缺血而坏死。

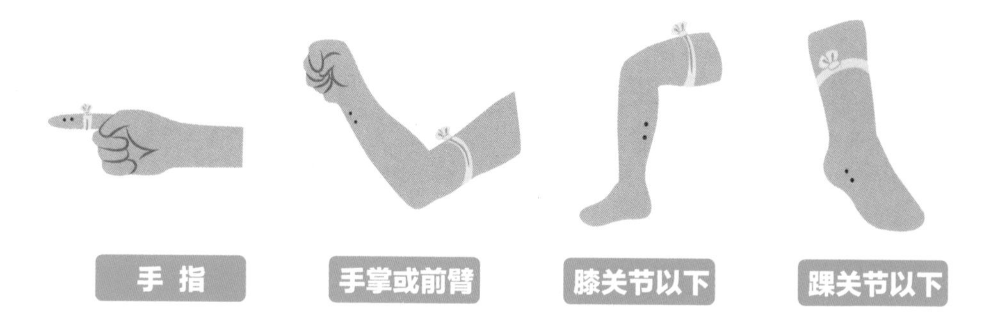

❸ 冲洗

用干净的清水、淡盐水或生理盐水冲洗伤口及周围皮肤，可由近心端向远心端（伤口中心）轻轻挤压伤口，排出毒液。但不提倡用嘴巴吸出毒液，因为毒素会通过口腔黏膜进入人体，使施救者也陷入危险。

④ 扩创排毒

用消毒干净的小刀、锐器把伤口牙痕处皮肤挑开，深达皮下，划一个“十”字（操作时不宜过深，注意避开血管和神经），有毒牙的应取出，加速排毒，切开后继续用清水冲洗。此方法在条件不具备的情况下不主张采用。

另外，可用打火机或点燃的火柴对准伤口位置直接烧灼，烧至伤口皮肤起水泡即可。该原理是高温能分解蛇毒中的蛋白，使其失去毒性。该法适合在咬伤30分钟内进行，时间越早，效果越好。

⑤ 就医

紧急处理完毕后宜尽快前往医院治疗，记住咬伤的毒蛇外观特征，以便医生对症治疗。

Snake Bites

When participating in outdoor activities, caution should be taken to prevent snake bites. Once accidentally bitten by a venomous snake, venoms will enter into the blood circulation and spread through the whole body within several hours, sometimes even a few minutes. So time is life.

(1) Call for help: Call 120 or other emergency number immediately to inquire about the nearest hospital that has antivenom drugs.
(2) Keep calm and restrict movement: Help the patient to lie down, if possible, keep the bite below the level of the patient’s heart. Create a loose splint to help restrict movement of the affected area. Remove jewelry and tight clothing, because the affected area may swell.
(3) Monitor the patient's vital signs: Temperature, pulse, rate of breathing, and blood pressure should be monitored.
(4) Do not drink caffeine or alcohol: These drinks could speed the absorption of the venom.
(5) Wash: Wash the wound and the surrounding skin with clean water, brackish water or soap water. Extrude the wound gently from the proximal part to the axifugal part to discharge poisonous blood.
(6) Medical treatment: After the emergency treatment at the scene, the patient should be sent to the hospital as soon as possible. Remember the snake species to give the doctors clue on appropriate solutions.

断桥残雪

CHAPTER

10/地震

Earthquake

地震是群灾之首，往往在我们毫无防备时突然降临。当地震发生在陆地上，会引起地面及建筑物崩塌、地面裂缝、山体滑坡等灾难；发生在海底或沿海地区，能引起海啸。学会地震时自救，是每个民众所必须掌握的逃生技能。

10.1 了解地震

地震开始发生的地点称为震源，震源正上方的地面称为震中，震后破坏最为严重的地区称为极震区，从震中到地面上任何一点的距离称为震中距。

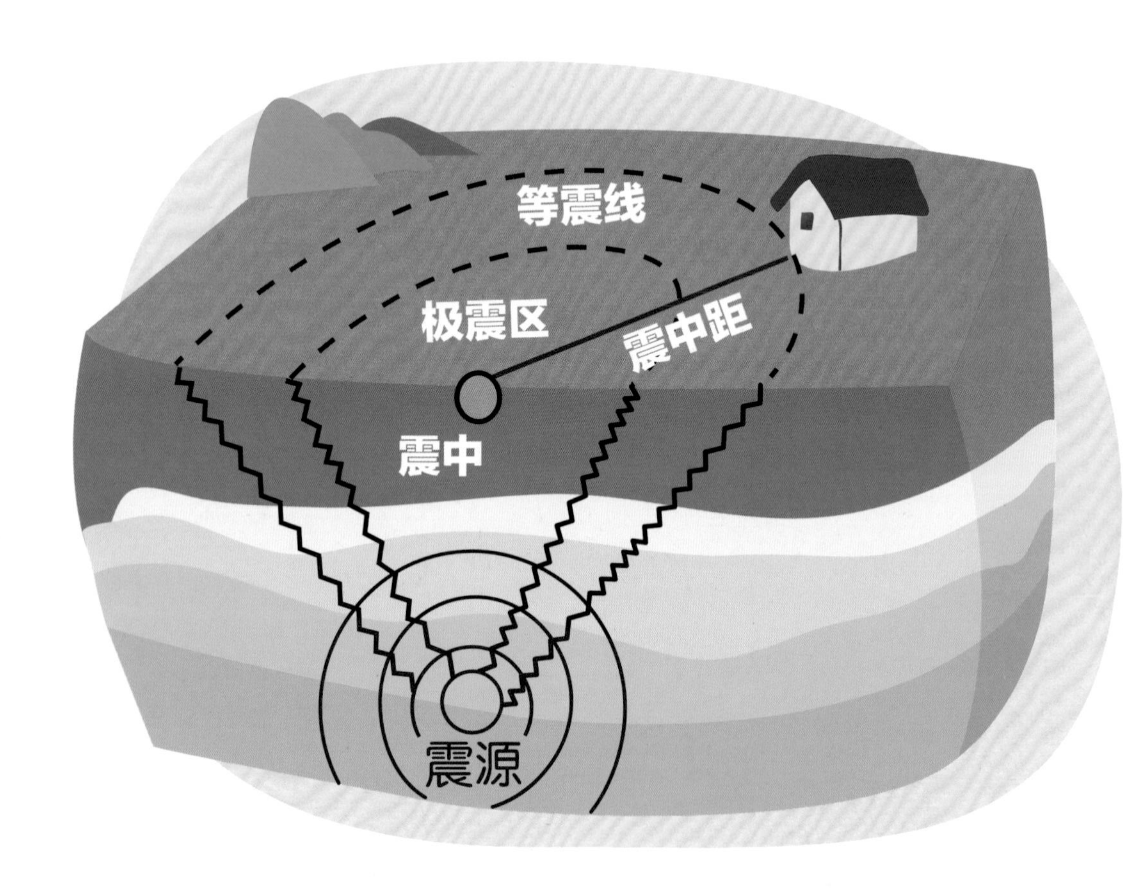

按震级大小可把地震划分为以下几类：

地震分类	地震震级大小	特点
弱震	＜3级	如果震源不是很浅，人们一般不易觉察
有感地震	3级≤震级≤4.5级	人们能够感觉到，但一般不会造成破坏
中强震	4.5级≤震级＜6级	可造成破坏，但破坏轻重还与震源深度、震中距等多种因素有关
强震	≥6级	
巨大地震	≥8级	

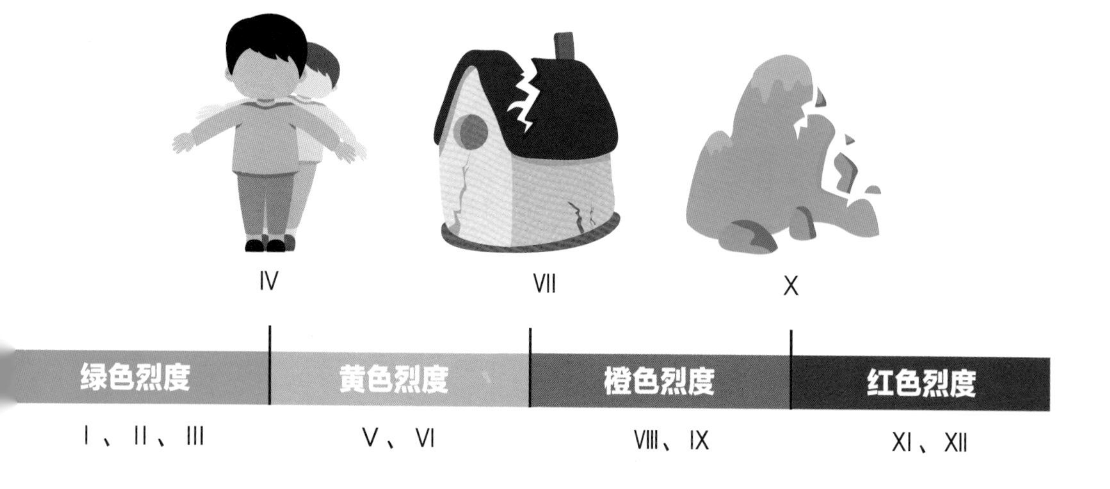

当地震发生时，大地振动是最直观的感受，一般在最初的12秒钟，你会从感到上下震动到左右晃动。12秒钟就是从地震发生到房屋倒塌的时间，也是我们的黄金逃生时间。

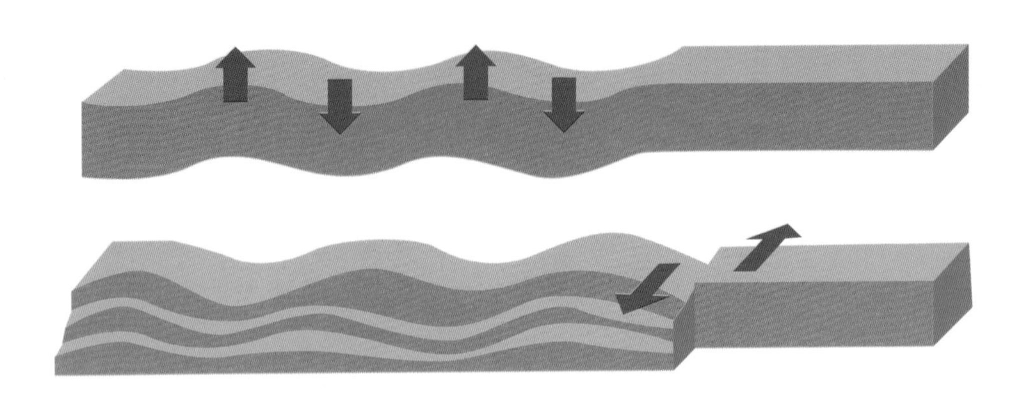

10.2 如此紧迫的时间，我们该怎么办

当地震发生时，无论你身在何处，都需保持镇静，在关键12秒钟内作出正确躲藏与逃生的抉择。

10.2.1 当你正在高楼层时

（1）就近躲避，蹲、坐或趴下，尽量蜷曲身体，用枕头、棉被、脸盆或坐垫等护好头颈部，或躲到床底或写字台下，因为它们的材质相对坚硬，有一定的抗压性能。但现在很多床底都无法藏身，那么也可紧贴床边，趴在地上，低头、闭眼、闭口、用鼻子呼吸，如可能用毛巾或其他物品遮住口鼻。

（2）地震时门窗可能变形，不要试图外逃。如你正处于门边时，可以先打开房门，切记不可从门窗处逃生或从楼上跳下，禁止使用电梯；地震过后迅速撤离时，也要走楼梯，记得一定要穿鞋子，以免脚受伤。

（3）远离外墙、门窗及镜子，也可躲在结实、不易倾倒、能掩护身体的物体旁，比如内墙墙角、紧挨墙根的家具旁、厨房、卫生间等，尽可能抓住身边牢固的物体，以防摔倒或身体移位。

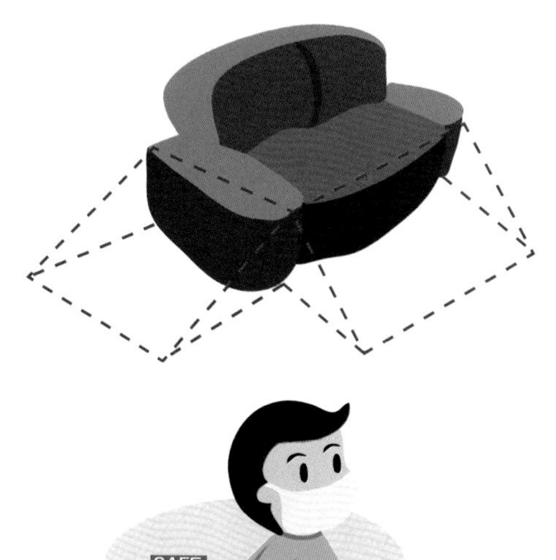

当房屋倒塌时，能躲在一个比自己高又比较结实的物体旁边，就等于找到了一个相对安全的能形成三角空间的支点。

（4）待震后，再逃到室外空旷、安全的地方。

（5）如你正在平房或楼房最低一层时，应充分利用12秒钟时间力争跑到室外，来不及跑时可按照高楼层内的方法进行躲避。

10.2.2 当你正在学校时

（1）躲到书桌底下，双手抱头，也可将书包盖住头部，防止被砸伤。

（2）不能站在书架旁、玻璃窗户旁，以防被压倒或玻璃碎片刺伤。

（3）听从老师指挥，有序撤离到学校操场等空旷处。

10.2.3 当你正在户（室）外时

若恰巧在比较开阔的地方，最好保持原地不动蹲下或趴下。

正在开车时，立即停车，可以先选择待在车内不动，视周围情况选择迅速逃向空旷处。

不能在公交站牌、天桥下、雨棚处躲避，也不要站在路灯旁边，避开高楼、大树、广告牌、橱窗、悬着高压电线的地方，注意高空坠物。

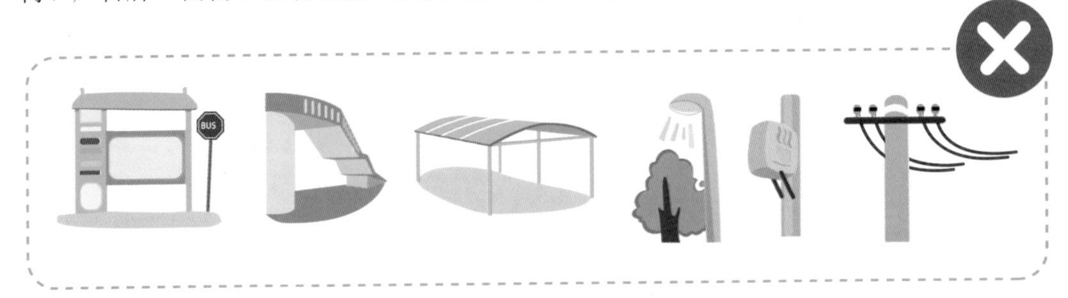

不要躲在隧道、地下通道内，以防地震时出口堵塞或隧道、地下通道坍塌。

不要急着跑回家，找寻家人。因为很多地震案例表明，在地震过程中，进入建筑物内，坠落物砸死砸伤的概率最大，你要相信他们在屋里也会做好应急保护措施。保护好自己，才有可能在地震结束后，加入到营救团体中，救助家人。

10.2.4 当你正在公共场所时

地震发生时的公共场所，最忌讳人群慌乱造成混乱场面或者踩踏事件，导致不必要的伤亡，所以每个人都要保持镇静。

在车站、商店、餐厅、医院等场所的人员，就地选择墙角、排椅、桌子下、床底等一切能够躲避处抱头躲藏；在剧院、体育场、体育馆等地时，就躲在排椅之间，不要乱跑、乱挤，观察等待时机。

处于地铁上的人员，切记不可强行开门，应按照地铁应急开门方式进行手动开门，注意听广播，按照广播指示有序撤离。

注意事项

- 在时间允许情况下，争取时间熄灭明火、关闭煤气和电源、把门打开（因为强震后有可能打不开门）。
- 地震时照明最好用手电筒，不要使用火柴、打火机等明火，因为可能有易燃易爆气体泄漏在空气中。
- 强烈地震过后多有余震，不要先急着回家。
- 如遇到与其他人一起被压埋在一个地方，一定要相互鼓励。
- 如一时之间找不到脱险办法，一定要保存体力，不要随意浪费力气，可以利用敲打墙壁、硬物、水管等向外界求救，尽量寻找水，延长生存时间，等待救援。

Earthquake

Earthquakes can cause severe damage and destruction; citizens should be prepared for a major earthquake. It is important to keep calm in the event of an earthquake. The first 12 seconds is crucial to make a right choice for survival.

1. When you are on a high rise building

If inside, limit your movements during an earthquake to a few steps to a nearby safe place. Crouch, sit or bend over, curl into the "turtle position". Protect the head and neck with pillows, quilts, basins or cushions. Take cover under a bed or desk. Bend your head, shut your eyes, and close your mouth, then breathe with the nose. Also if possible, cover your face with a towel or other fabric to prevent coughing.

Doors could distort and windows could shatter during the earthquake, so do not try to go outside, this is unless you are near a door that you can open. Do not use elevators in any situation. Shoes should be worn to protect the feet.

Stay away from external walls, windows, doors and mirrors. Take cover under solid objects, such as the corner of an internal wall, furniture next to the foot of the wall, the kitchen, or a toilet. Try to grasp a firm object to avoid falling down.

Stay indoors until the shaking has stopped, and then go to a safe, wide area outside.

If you are in the first floor, try to escape within 12 seconds.

2. When you are at school

Take cover under the desk; cover your head with hands or schoolbags.

Do not stand beside bookshelves or glass windows.

Follow the teacher's instructions and evacuate calmly to the courts or other open spaces.

3. When you are outdoors

If you are in an open space, it is best to stay in place. Go prone or crouch until the shaking stops.

If you are driving a vehicle, pull over to a clear location, and stay there with your seatbelt fastened until the shaking has stopped. Check around and make a choice whether you should escape quickly to an open space.

Do not take shelter under a bus stop, overpass, or canopy; stay clear of buildings, trees, streetlights, advertising board and electricity lines, and anything that could fall down.

Do not take shelter in a tunnel or underground passage, as they can easily get blocked with falling objects during an earthquake.

4. When you are in a public area

Everyone should keep calm.

If you are in a bus, railway station, shopping mall, restaurant or hospital etc., cover head with your hands and hide under tables, benches, or in the corner of a wall. When you are in a concert, stadium or gymnasium, shelter between benches and do not run in a rush, do not push and shove, and wait for a chance for escape.

If you are in the subway, do not force the door open. Follow the instructions to open the door manually and evacuate with order following the instructions from the broadcasting.

宝石流霞

CHAPTER

11 / 火灾

Fire

火灾是日常生活中较常见的灾难之一，在任何时候、任何地点，都有可能发生。一个粗心未熄灭的烟蒂、一根尚未燃尽的火柴，在一定环境下，都有可能引发一场火灾，有时大自然高温下也能聚集热量燃烧周围物品，形成熊熊大火。

11.1 如何正确使用灭火器

避免火灾，最好的方法是预防。目前城市各建筑楼、公共场所、交通工具等处都配置相应的灭火工具，所以我们首先要学会的是使用灭火器的方法。

常用灭火器种类：二氧化碳灭火器、干粉灭火器、泡沫灭火器、1211灭火器（卤代烷灭火器）等，不同灭火器内装的成分不同，使用的对象也不同，使用时要先看清楚。

泡沫灭火器

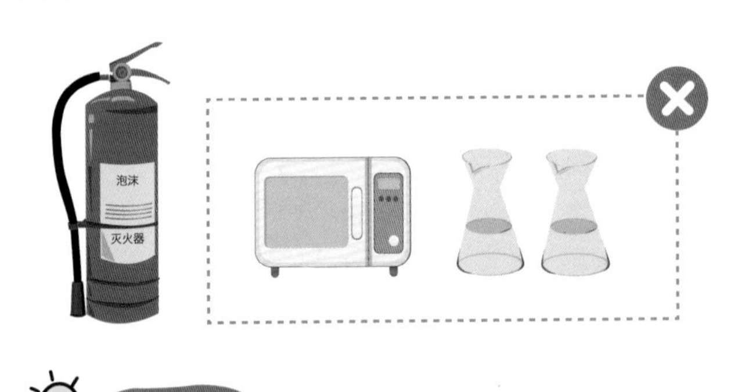

温馨提示

酒精引发火灾和带电电器火灾不能用泡沫灭火器灭火。

灭火器使用方法

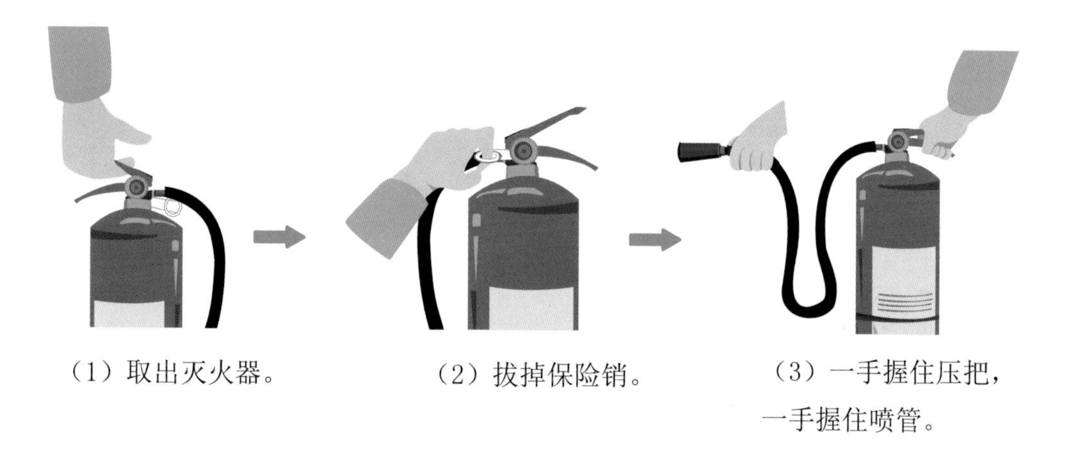

（1）取出灭火器。　（2）拔掉保险销。　（3）一手握住压把，一手握住喷管。

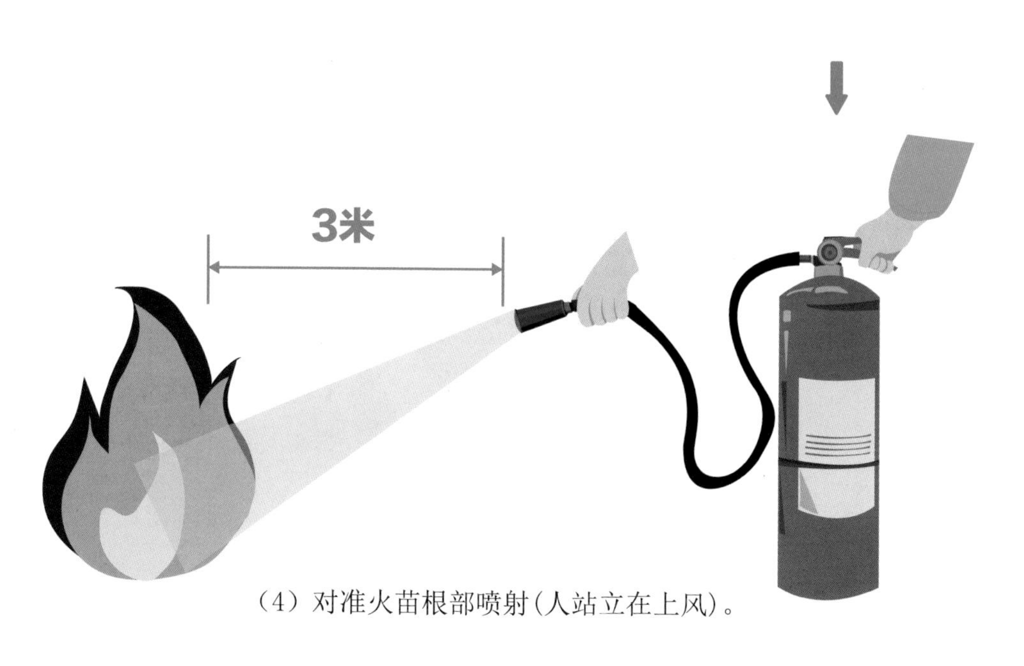

（4）对准火苗根部喷射(人站立在上风)。

11.2 当火灾来临时，如何自救

1. 任何灾难的发生，都需保持镇静，不惊慌，不绝望，不盲目逃生。
2. 立刻拨打“119”报警，清楚告知以下内容：

- 什么地方着火了（起火地址，包括所在区域、路段、胡同、小区名字、几幢几单元几室或乡村地址、门牌号、户主名字，附近有无明显标志等）？
- 什么东西着火了？有无危险品？
- 火势大小如何？烟雾大小及颜色？
- 有无人员被困？
- 留清楚你的姓名和手机号码。
- 等待对方明确说明后方可挂电话。
- 在路口明显处等候消防车，指引其去火场道路，然后撤离到安全地带。

11.2.1 处于室内时

（1）身陷火场时，小火使用灭火器，大火难灭时要迅速逃离。

在逃生中，不要迟疑，不要因顾及个人财产而返回房间内取物，生命才是第一位的。

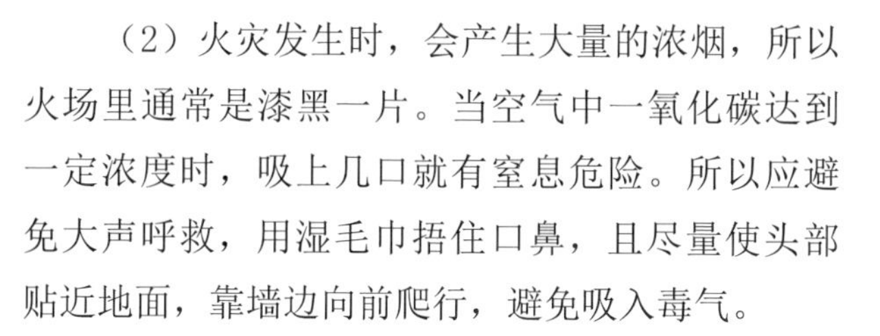

（2）火灾发生时，会产生大量的浓烟，所以火场里通常是漆黑一片。当空气中一氧化碳达到一定浓度时，吸上几口就有窒息危险。所以应避免大声呼救，用湿毛巾捂住口鼻，且尽量使头部贴近地面，靠墙边向前爬行，避免吸入毒气。

千万不要往柜子里或床底下钻，也不要躲藏在角落里，更不要盲目往火场跑。

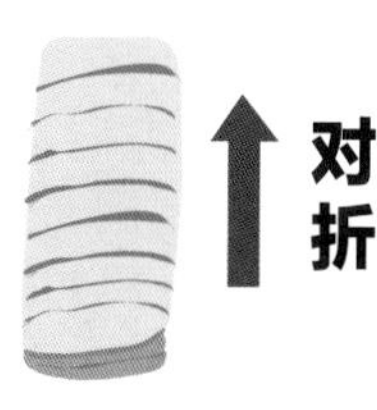

如图所示，将打湿的长方形毛巾对折3次，叠成8层

也可在身上浇冷水或裹上浸湿的衣物、棉被、毯子等向安全出口方向冲出去，朝火灾上风向逃生。

朝着火灾上风向逃生

（3）逃生时一定要走安全通道，不可跳窗或乘坐电梯。

（4）由于火势向上蔓延，当发生火灾的楼层在自己所处的楼层之上时，应迅速向楼下跑。如下楼通道受火势阻断时，可利用疏散楼梯、阳台等逃生自救。

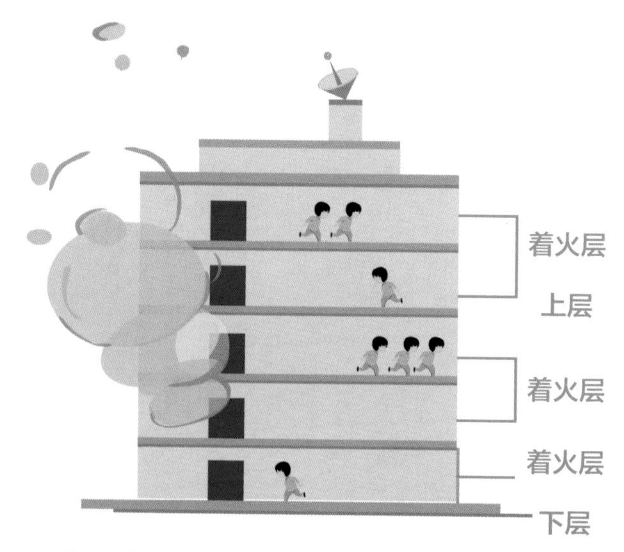

高层建筑内迅速判断火源，远离火源处逃生

可用绳子或把床单、窗帘布撕成条状连成绳索，紧拴在窗框、管道、铁栏杆等固定物上，用毛巾、布条等保护手心，顺绳滑下，或下到未着火的楼层来脱离险境。

如在二楼，万不得已情况下，可先向地面扔棉被、床垫等，跳楼逃生。

（5）假如你所在的地方已被大火封闭，或用手摸房门时感到烫手，此时，应该暂时退入房内。一旦开门，大火与浓烟势必迎面而来。关闭所有通向火区的门窗，用浸湿的被褥、衣物等堵塞门窗缝，并向门上泼水降温。

积极向外寻找救援，用打手电筒、挥舞色彩明亮的衣物、呼叫等方式向窗外发送求救信号。

如身上着火，采用滚动方式灭火。

11.2.2 处于公共场所时

（1）立即按响警报铃。

教你正确使用消火栓箱

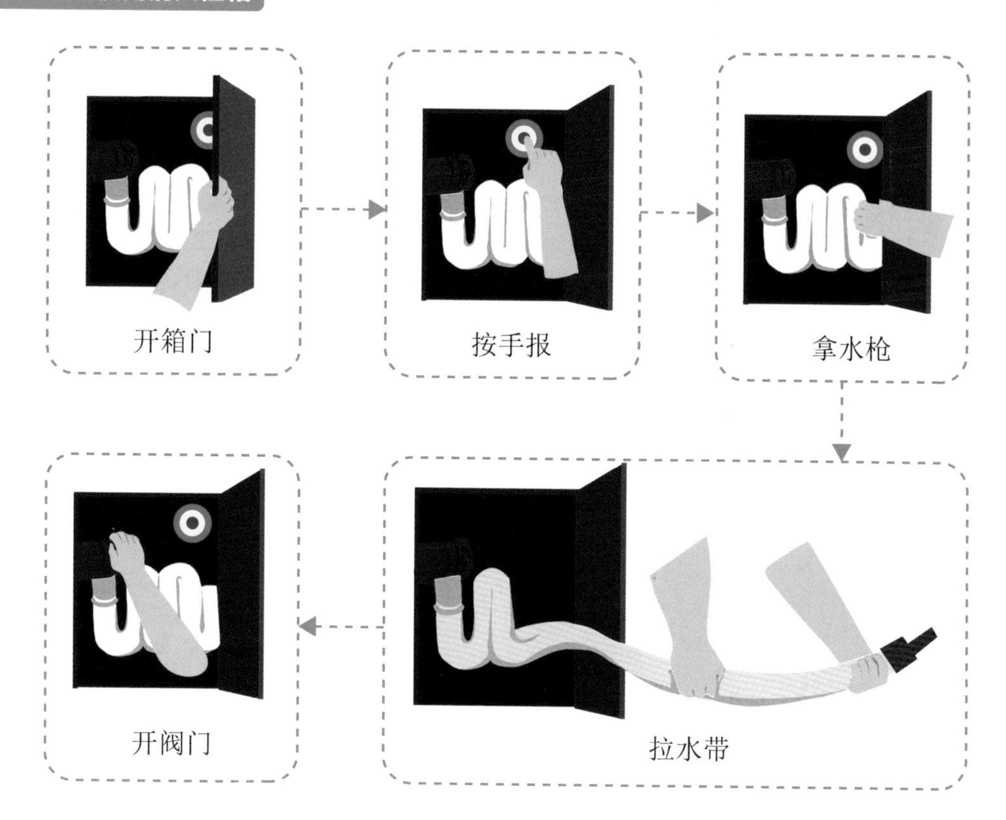

（2）认真听广播，听从指挥和引导，从应急通道逃生。

（3）身处险境，除了自我逃生之外，在能力范围内要积极救助老、弱、病、残、孕等人逃生。

11.2.3 我们能做哪些预防措施

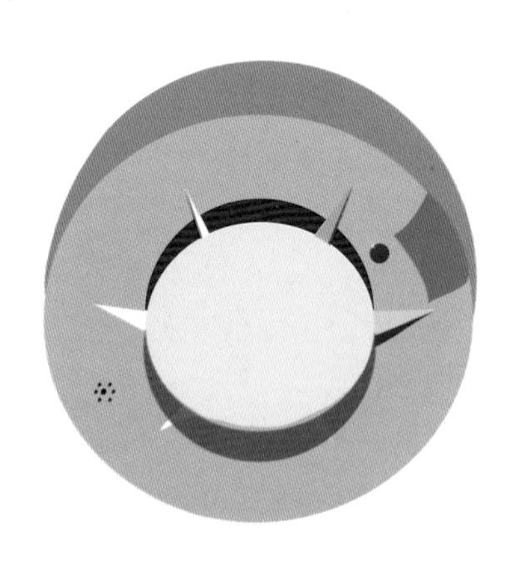

可在家中安装住宅用火警警报器，从而能够提早发现火灾。

最好能配备灭火器、灭火毯、简易防毒面罩、手电筒、静力阻燃绳等，以备不时之需。

制订家庭逃生计划图，画出家里的平面图，标记逃生出口，注明逃生火场地点、路线及屋外集合地点，贴在进门的地方。

家里最好不要安装固定防盗窗，以防逃生时被困。

切勿在走廊、楼梯口堆放杂物，保证通道和安全出口的通畅，防止逃生时摔倒。

Fire

Fire spreads very quickly. Activate the fire alarm and call for help are the first priority. It is important to grasp some survival skills.

1. If you are indoors as the fire occurs

(1) If the fire was small, you can use the fire extinguishers to put out the fire if you are trained; if the fire was difficult to put out, you should try to escape as soon as possible.

(2) Avoid shouting, cover your nose with wet towel, put your head close to the ground, and crawl forward along the wall, and avoid breathing the toxic gases. Do not go into the cupboard, under the bed or in a corner, also do not run into the fire.You can douse yourself with cold water, or cover yourself with wet clothing, quilts, blankets, and then run upwind.

(3) You should escape using the stairs.

(4) If the fire is on a higher level, you should run downstairs rapidly. If the stairs are on fire or blocked, use the fire-escape stairs or balcony.

(5) If the place you are staying was blocked by the fire, or the door of the room was burning, you should return to the room, close the windows and doors facing the fire, fill the gap of door and window with wet quilt or clothing, and pour water for cooling. Actively seek help, and call for help from the window with a flashlight, brightly colored clothes or shout.

2. If you are in a public area

Ring the alarm bell immediately, and then run away through the emergency escape route following the instructions.

平湖秋月

CHAPTER

12 台风

Typhoon

夏季是台风的频发季节，它是发生在热带洋面上的猛烈风暴，是我们无法避免的一种天气现象，波及范围广，来势凶猛，破坏性极大，是最具危险性的灾害之一。

12.1 台风发生的规律及特点

- 有季节性，台风（包括热带风暴）一般发生在夏秋之间，最早发生在5月初，最迟发生在11月。
- 中心登陆地点难以准确预报。
- 风向具有旋转性，台风的风向时有变化，常出人预料。
- 常伴有大暴雨、大海潮、大海啸。
- 强台风发生时，人力不可抗拒，易造成人员伤亡。

侧 台风的侧视图

俯 台风的鸟瞰图

12.2 台风现场的危害

台风的破坏力主要由强风、暴雨和风暴潮三个因素引起，从而直接引起建筑物、公共设施等的破坏，间接对人类造成伤害。伤者通常为户外活动的人员，以坠落伤、塌房致伤、砸伤、割刺伤、车祸、电击伤、淹溺等导致的损伤居多，受伤部位多以头部外伤、皮肤刮伤、四肢骨折、颅脑损伤、胸腹部损伤等为主。

气象台根据台风可能产生的影响，在预报时采用“消息”“警报”“紧急警报”三种形式告诉民众。同时，按照台风可能造成的影响程度，从轻到重以“蓝、黄、橙、红”四种颜色作为信号。

蓝色台风预警　一般

6~7级　树木会摇晃

黄色台风预警　较重

树叶满天飞

橙色台风预警　严重

10~11级　树木被吹断

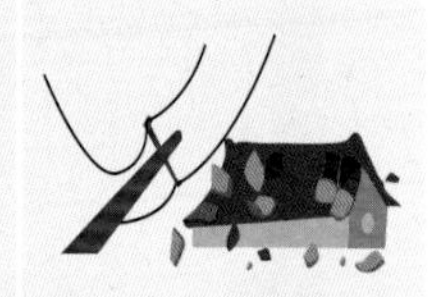

红色台风预警　特别严重

12~13级　吹倒电线杆屋顶砖也掉了

14~15级　强台风，灾难性损害

16级　超强台风，更严重灾难性损害

12.3 台风来之前，可以做哪些准备

及时收看台风相关报道，弄清楚自己所处的区域是否是台风袭击的危险区域。

最好能准备好收音机与干电池，以防断电后无法及时了解最新情况。

检查门窗是否牢靠，取下或固定好容易被风吹动的招牌等；将房屋外所有悬挂物取下，尤其是顶楼和阳台上的花盆与杂物；清理阳台排水沟。

备好手电筒或应急灯、手机电池或充电宝、蜡烛、打火机或火柴；饮用水、食物等生活用品。

住在低洼地区、山坡边或危旧房屋的民众，应尽早撤离并转移到高处或安全地带。

取消一切登山等户外活动，千万别进行海边游泳、观浪、钓鱼、戏水等不恰当的活动。

检查电路、煤气等设施是否安全。
把汽车开到高处，以免淹水。

12.3.1 台风来了怎么办

尽可能躲避在屋内，不要外出活动。若人恰巧在室外，应躲避在坚固的墙壁下，也就是背着风的一面。标准姿势是抱住头蹲在地下，尽可能缩小身体受风的面积。

如果必须外出，最好穿雨靴，防触电；千万不要在河、湖、海的堤坝上或桥上行走；要穿上颜色鲜艳、紧身合体的衣服，行走时弯腰把身体缩成一团，以减少受风面积，尽可能避开地下通道等易积水地区。

关紧门窗，必要时加钉木板，进行加固措施。

在窗玻璃上用胶布贴成“米”字图形，以防玻璃窗破碎。

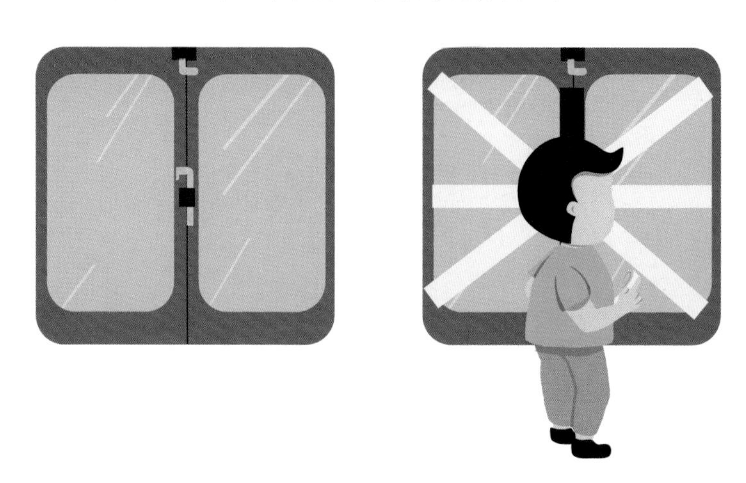

在一楼生活的民众，有能力者应堆好沙包，以防进水。

如遇雷雨大风，应及时关闭煤气，关闭正在使用的家用电器，并拔出插头。如家中不慎进水，应立即切断总电源。

台风期间不要开车，不得已在外开车时，应减速慢行，积水超过轮胎一半时，不可强行通过。

如行驶途中车辆遇台风袭击，应立即将车开到地下停车场或停靠在路边安全地方，迅速下车，找安全建筑物躲避，千万别存躲在汽车内躲避台风的侥幸心理，台风来时，汽车不足以与台风相抗衡。

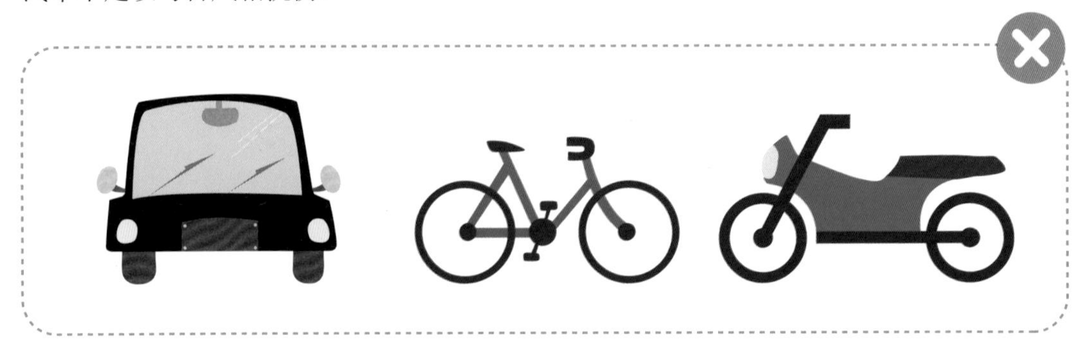

骑自行车、电动自行车或摩托车在风中受到的冲力，比步行更大，车头易失控，所以台风直接影响期间，最好勿骑车。

谨防高空坠物。室外一定要注意躲避以下设施：广告牌、路灯、树木、电线杆，还有铁塔、棚架、临时建筑、在建工程的施工用具等，防止这些东西在强风下倒塌，砸下伤人。

出海船只迅速返回港口，进入避风港。

台风过后需要注意环境卫生和食物、水的安全。看到落地电线，不要靠近，也不可用手触摸，以防触电，可以先帮忙竖起一块警示标志，然后再拨打电力热线报修电话。

Typhoon

Summer is the season of typhoons, which has the characteristics of being destructive. Strong wind, heavy rain and storm tide are three main causes of destruction during typhoon. Injury is common, such as fall injury, cut injury and electrical injury. Measures should be taken to prevent injury.

1. What can we do before a typhoon is coming?

(1) Watch any typhoon forecasting, and prepare a radio and batteries for receiving new information in the case of a power cut.

(2) Check if doors and windows are tightly closed, and remove or fix any objects that are easily being blown down, such as a sign board; take down anything that is hanging outside the house, especially flowerpots and other things on the top floor or balcony; clear up the drainage ditch of the balcony.

(3) You need to get these things ready: a flashlight or emergency lights, batteries for your mobile phone, a power bank, candles, lighters or matches, drinking water, food and other essential things.

(4) The people who live in low-lying area, hillsides or dilapidated houses, should move to a higher and safer area.

(5) Cancel all outdoor activities, such as mountaineering.

(6) Check the power circuit, gas and other facilities.

(7) Park your car on a high altitude to avoid being flooded.

2. As the typhoon comes, what should we do?

(1) Stay indoors.

(2) If you have to go outside, you'd better wear rain boots to prevent any electric shocks; do not walk on the bank of a river, lake or sea, especially not on a bridge; wear fit clothes that are bright in color, bend your body to reduce the area the wind can effect; try to avoid those places which may easily collect water as much as you can.

(3) Stick adhesive tape onto the window glass in the shape of "米" to prevent the glass from breaking.

(4) People who live on the first floor should pile the sandbags at the door to prevent any water from flowing in.

(5) If there is a storm, thunder or strong wind, please turn off the gas, turn off any electrical equipment and pull the plug out immediately.

(6) As the typhoon comes, do not drive, cycle or engage in any other outside activities, as to avoid falling objects.

(7) After the typhoon, we should pay attention to the environment's hygiene, including the safety of food and water. If there is wire on the ground in view, do not get close or touch it, you can set up a warning sign, and then dial the repair hotline of the electric power company.

龙井问茶

CHAPTER

13/踩踏事件

Stampede

踩踏是人为的突发事件，一般发生在大型活动中，因人群数量过多或过于拥挤，导致大部分人走路或站立不稳后摔倒，不能及时爬起，被其他人踩在脚下或压在身体下，最终导致短时间内人群发生连锁倒地。

预防踩踏事件发生的最好办法，是大家提高安全避险与安全防范意识，在进入任何拥挤场所时，留意并记住出口和紧急通道。

13.1 哪些情况下容易造成踩踏事件

在空间有限、人群又相对集中的场所内，大型活动或节日时，比如球场、商场、影院、酒吧、夜总会等，宗教朝圣的仪式上、音乐节现场等。

当人受到一定的惊吓，比如火车或地铁上的爆炸声、枪声、火灾发生时，因慌乱逃生也很容易发生踩踏事件。

13.2 踩踏事件发生时，如你恰巧身在其中，请按照以下方法，尽量减少伤害

（1）人流拥挤时，不要慌乱、保持冷静，尽量顺着人流方向移动，不要反向走，也不要超赶往前挤，要当心被绊倒。

（2）在人流中，身体不要倾斜失去重心，假使鞋子被踩掉，也不要弯腰捡鞋子、系鞋带。

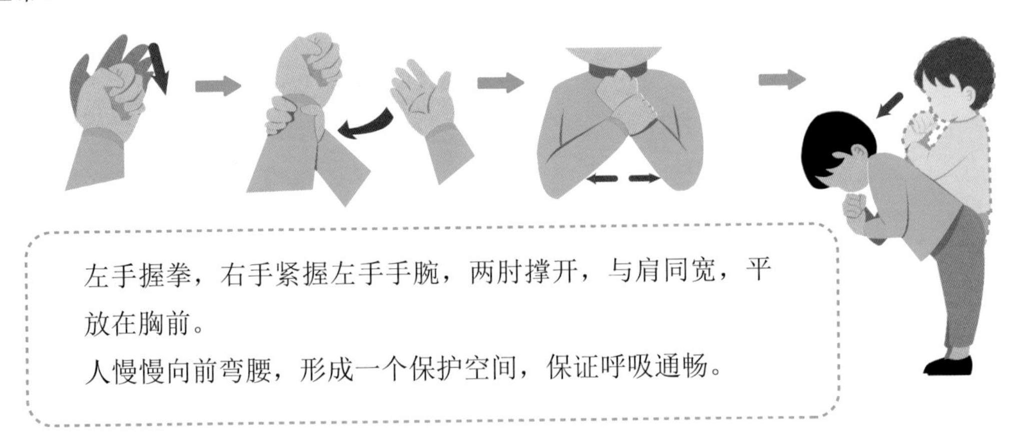

左手握拳，右手紧握左手手腕，两肘撑开，与肩同宽，平放在胸前。
人慢慢向前弯腰，形成一个保护空间，保证呼吸通畅。

（3）寻找附近有无牢固物体，如有可先尽快抓住或慢慢停住。遇到台阶或扶梯时，尽量抓牢扶手。

（4）当发现自己前面有人突然摔倒了，停下脚步的同时，要大声呼喊，告知后面人不要再继续前行。

（5）一旦被人群推倒在地上，要设法让身体靠近墙角。

身体缩成一团，双手在颈后紧扣，保护颈部和后脑；两肘向前，护住太阳穴。

13.3 如何救助他人

人类与生俱来的逃生本能，往往可能导致在惊慌与奔跑中造成人群的巨大伤害，受伤人数多、伤情重、现场急救处理也会比较复杂。踩踏事故中比较容易发生的有骨折和挤压伤。

- 对于骨折伤，不能随意移动伤肢，可以就近取材固定伤肢，按照固定技术进行处理。
- 对于挤压伤，应尽快解除对身体的压迫，固定好伤处，禁止搬动伤员，按摩、热敷、抬高身体等，应原地等待救援。
- 对没有呼吸的伤者，应按照心肺复苏方法立即进行抢救，直到医护人员到达。

Stampede

Stampedes can occur during professional sporting and music events, where large crowds gather in small areas with little crowd controls. Caution should be taken to avoid any crushing.

In the case of a stampede emergency, follow the instructions below to decrease the chances of injury.

Help yourself

(1) If you are stuck in the middle of a big crowd, keep calm and do not panic. Follow the stream of the people, not the reverse, and walk slowly, avoid pushing and stumbling.

(2) Keep the body stable and do not lean forward or backward. Do not bend down in any situation, even if you lose a shoe.

(3) The left hand makes a fist, and the right hand holds the wrist of the left hand tightly. Open your elbow rights under the shoulders, and put hands on the chest.

(4) Lean the body forward slightly, creating a protected space for breathing.

(5) In the case of stairs or steps, hold the handrail tightly.

(6) If someone falls down in front of you, stop and shout out for others around you to notice and stop.

(7) If you fall, get up quickly. If you cannot get up, keep moving by crawling in the same direction of the crowd, or if that is not possible, try to move to the corner of a wall, then cover your head with your arms and curl up into the fetal position (do not lay on your stomach or back, as this dangerously exposes your lungs).

How to help others

Bone fracture and getting crushed are the most common injuries during the stampede.

(1) If there is potential bone fracture, do not move the injured part of body but fix its position.

(2) If there is potential crush injury, remove the pressure to the body as soon as possible. Secure the wound and do not move it around.

(3) In case of the absence of breathing, follow the steps of CPR until the arrival of health-care providers.

虎跑梦泉

CHAPTER

14 / 危险品爆炸

Explosion Hazard

危险品爆炸的场景我们在电视里都见到过，现场产生巨大的爆炸声，火光冲天，不仅造成重大财产的损失，而且让身处周围的人们都很难幸免。当消防员们在为我们奋不顾身时，我们是否能为自己与他人做点什么呢？

14.1 首先，我们需要了解哪些是危险物品

它是指易燃易爆、危险化学、放射性等能够危及人身安全和财产安全的物品。

危险品爆炸后会产生火光、爆炸冲击波、浓烟，容易对人造成烧伤、致残、中毒、窒息等。有些危险品还具有放射性，人体接触后会造成急性或慢性的放射性疾病，但不是所有爆炸带出的气体都是危险的，其取决于爆炸物所含成分、数量、距离等。

认清危险物标识

ASBESTOS
石棉
Do not Inhale Dust
切勿吸入石棉尘埃

14.2 面对突如其来的危险品爆炸，该如何自救

14.2.1 公众场所爆炸

1 立刻趴下

当爆炸发生时，保持镇静，立刻用手护住头，背朝冲击波传来的方向趴下或蹲下，别急着跑，这样不但可以最大程度降低爆炸冲击波的伤害，也可防止吸入过多的有毒烟雾。

2 向上风向快速撤离

观察周围环境，确认短时间内不会发生二次爆炸后，向最近的安全地点逃离，爆炸现场如有浓烟，应往上风向弯腰逃生。用打湿的毛巾或布蒙住口鼻，不要叫喊，减少烟气吸入。尽量不乘坐电梯。

③ 寻找掩护

当爆炸发生后，不要待在密闭空间里，应选择在开阔的环境或能够有效阻挡、反射爆炸冲击波的地方，如建筑物、土围墙、汽车、家具等物体背后，但尽量远离门窗、柱子、玻璃、管道口、沟渠等位置。

④ 脱险求救

当确认自己已处于较为安全的地点后，立即拨打“120”“110”“119”等急救报警呼救，并向救援人员准确说明爆炸地点和时间。

14.2.2 室内场所燃气爆炸

（1）如闻到浓重煤气味道但还未起明火时，立即关闭阀门开窗通气，然后撤离到安全地点。

（2）如遇明火，立即往爆炸点上风口移动，禁止使用电梯；切记不可开关电器、不用打火机、远离气味浓重的地方再打电话报警。

（3）逃生过程中不要从爆炸点附近经过，寻找其他通道离开，切记不可返回拿贵重物品，生命安全应放在第一位。

14.2.3 公交、地铁爆炸

- 闻到焦糊等异味，立刻提醒司机停车检查。
- 听从司机指挥，切记不要拥挤，有序下车。

- 若车内已起烟雾，可根据公交车内的指示使用生命锤、高跟鞋等物品将门窗玻璃敲碎逃生；按照地铁应急提示方法，手动打开地铁自动门。

- 保持重心，抓住一切可抓住的扶手，不要弯腰、提鞋，以防止踩踏事件的发生。

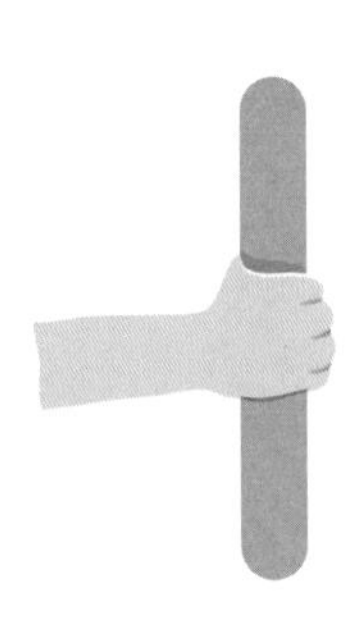

注意事项

- 如有专业人员到达现场指挥救援，一切行动听从指挥，不要贸然进入火场救人。
- 万一身体着火，应立即脱掉着火衣服，远离火源；若衣服无法脱下，就地打滚灭火；切勿奔跑，这样会让火势更旺。
- 对于烫伤，不要自行涂药，应用流动清水冲洗10分钟降温，然后用清洁布类覆盖伤处，保护创面。
- 发现呼之不应者，安置其侧身躺地，保持呼吸道通畅；心搏骤停者，应立即进行心肺复苏。
- 流血伤口，按照包扎止血法或指压止血法进行处理。
- 请记住关键一点：财物身外物，生命最宝贵。

Explosion Hazard

Explosion hazard is becoming more frequent in recent years. Citizens should know how to react at first notice.

1. Explosions in a public place

(1) Bend over immediately: If you are in the area that was impacted directly, keep calm, protect the head with hands, and turn your back towards the blast wave.
(2) Evacuate ASAP towards the upwind direction: Check around quickly, and be sure there are no obvious explosions within a short time; bend over and run towards the upwind direction. If there is heavy smoke, cover the mouth and nose with a wet towel. Do not shout loudly to avoid inhaling toxic smoke. Do not use the elevator.
(3) Seek for shelter: Stay in an open space or seek for shelter, such as buildings, walls, cars, furniture, which could block and absorb the explosive wave. Stay away from doors, windows, pillars, glasses, pipe outlets, canals or ditches.
(4) Ask for help: Call for help ASAP, and tell the rescuer the location and time of explosion.

2. Indoors gas explosion

When there is a fire, move to the upwind direction of the explosion. Do not use the elevator. Do not turn on the light or use any lighters. Call for help when you are far away from the strong smell.

3. Public transport explosion

(1) Follow the driver's instruction. Do not push and rush. Evacuate the vehicle orderly.

(2) If there is smoke in the vehicle, break the window immediately. Use the emergency door-opening device when in the subway.

(3) Keep your center of gravity, grasp all nearby handrails. Do not bend over to wear shoes as to avoid being trampled.

九溪烟树

CHAPTER

15 / 中暑

Heatstroke

中暑，民间称为“发痧”，是指长时间受阳光直接暴晒、或在高温环境内进行体力活动，人体无法正常调节体温后发生的一系列急性疾病，是夏季常见的急症，发病突然，严重者可并发多脏器功能障碍。

15.1 人在哪些情况下容易中暑

- 在温度＞35℃的露天工作或活动，没有做好足够的防暑降温措施。
- 在温度接近≤35℃且空气中湿度较高的室内工作，通风差、散热困难。
- 身体素质较弱、肥胖、先天性汗腺缺乏症等人群。
- 在极度疲劳、睡眠不足、穿紧身不透气衣裤、挨饿缺水、大量流汗等情况下，再健康的人也会中暑。

人体排汗散热困难，水、电解质损失过多。

体温调节中枢功能紊乱

15.2 识别三种中暑症状

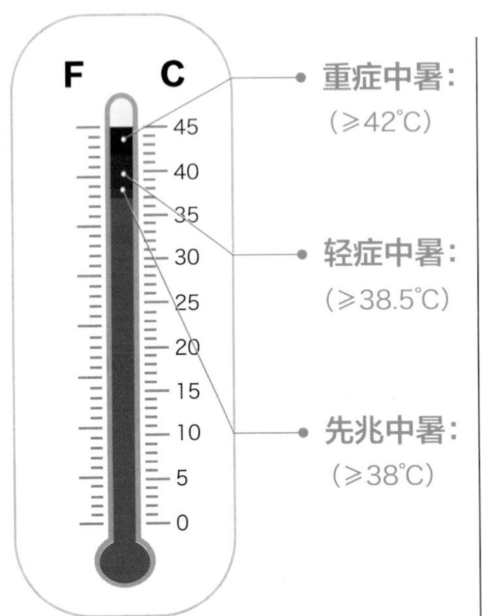

除轻症中暑的症状外，出现突然昏倒或抽搐、烦躁不安、高热、痉挛、晕厥和昏迷等症状，体温最高能升高到42℃以上。

除有先兆中暑的症状加重外，伴有面色潮红、皮肤骤热、大量出汗，恶心、呕吐、脉搏加快等表现，体温升高至38. 5℃以上。

在高温环境下劳动工作一定时间后，出现大汗、全身疲乏、口渴、头昏眼花、胸闷、心悸、恶心、注意力不集中、体温正常或略升高但不超过38℃的症状。

15.3 现场救护原则

❶ 环境

迅速将中暑者搬离到阴凉通风处或20—25摄氏度房间内平躺休息，下肢抬高15—30厘米。解开衣领、裤带或脱掉外套，保持呼吸通畅，若衣服被汗水湿透应立即更换衣服。

❷ 评估病情

先兆中暑者，喝0.1%凉盐水或清凉含盐饮料，短时间休息后就能恢复正常。

轻症中暑者，除了补充水分和盐分外，采取以下降温措施，一般3—4小时可恢复正常。

重症中暑者，进行以上处理的同时拨打“120”，有条件的需立即送到附近医院进行救治。在降温处理时或护送途中若发现患者没有意识（失去知觉），可指掐人中、合谷等穴位，如仍不醒来，或呼吸停止，应立即进行心肺复苏。

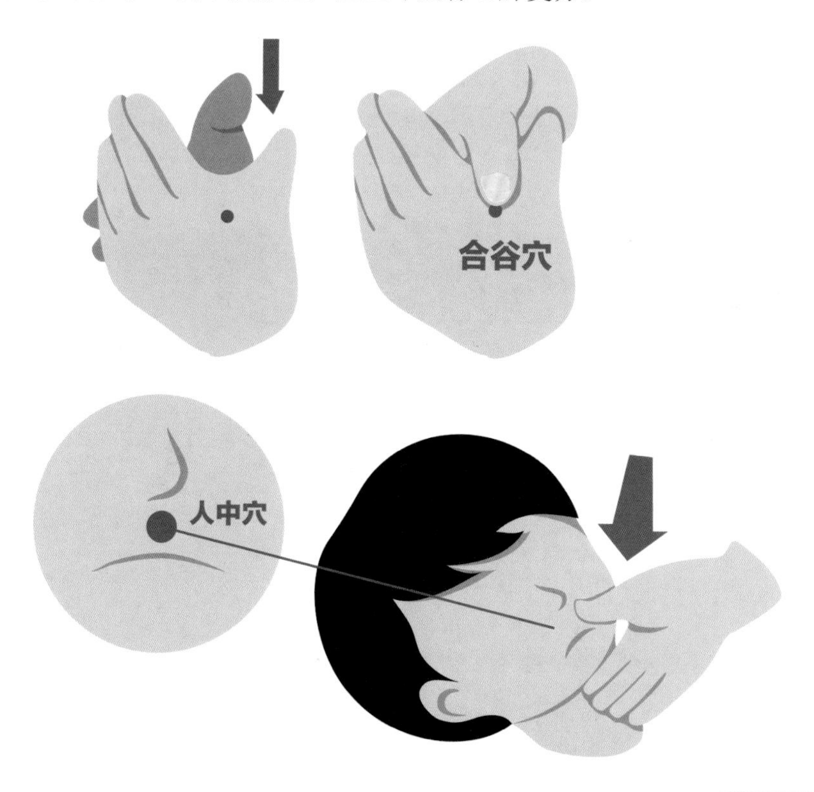

③ 降温

轻症中暑者可反复用冷（冰）水或酒精（30%—40%的浓度）擦身、额头、颈、腋下、大腿根部等大血管处放置冷毛巾或冰袋，使用扇子或电风扇加速散热。

注意事项

> 要注意及时更换冷毛巾或冰袋，尽量避免同一部位长时间接触，以防冻伤。当中暑者体温降到38℃以下时，停止一切擦身或冷敷等降温措施。

喝一些含盐冰水或饮料，不要急于补充大量水分，以免引起恶心呕吐。

刮痧法：用光滑平整的汤匙蘸食油或清水，刮背脊两侧、颈部、胸肋间隙、肩臂、胸窝及腘窝等处，刮至皮肤出现紫红色为止。

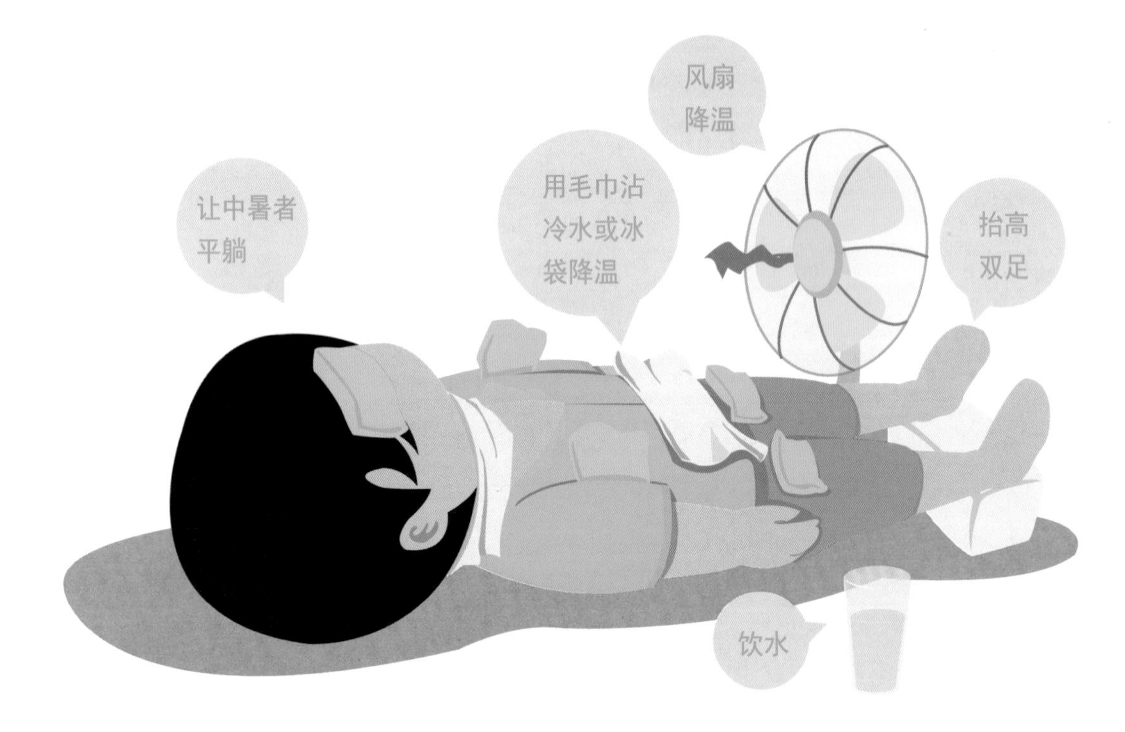

15.4 预防中暑

1. 避免长时间在高温及潮湿的环境下工作和运动。

2. 在露天劳动或外出时，穿较浅色、宽松透气的衣服，需要时可戴太阳帽或撑遮阳伞，做好防晒措施。备好防暑药品（人丹、藿香正气水、清凉油、十滴水等）及含盐分的饮料（浓度为0.4%—2%，即水2—5升+盐20克）。

3. 老年人、孕妇、患有高血压、心血管疾病、排汗功能障碍等疾病的人群，高温季节应尽量减少外出，尤其是在10—16点这段时间。

4. 饮食上建议少吃油腻辛辣食物，多吃新鲜蔬菜、水果，如番茄、绿豆汤、酸梅汤等，多喝水。

5. 科学作息时间，保证充足睡眠，也是预防中暑的好方法。

Heatstroke

Heatstroke, also known as sunstroke, is considered a medical emergency, caused by failure of the "thermostat" in the brain, which regulate body temperature. It is usually due to a prolonged exposure to heat or a high fever. If someone is experiencing symptoms, such as headache, dizziness, discomfort, chest tightness, palpitation, nausea, confusion, rapid deterioration in the level of response, hot, flushed and dry skin, body temperature above 40 ℃, these are the symptoms of heat stroke. In the case of heatstroke, cooling should begin immediately until the arrival of paramedics.

First aid for heatstroke at the scene

(1) Environment.

Quickly move the patient to an air-conditioned environment , or at least a cool, ventilated, shaded area or a room at the temperature of 20~25℃. Lie down to have a rest, with the lower limbs elevated up to 15~30 centimeters, making sure the patient can breathe freely.

(2) Evaluating the patient's condition.

If the patient is unconscious, press the phitrum and Hegu point, etc.. If the patient is still unresponsive, open the airway and check breathing. If there is no normal breathing or pulse, the rescuer should start CPR immediately.

(3) Cooling processes

① Remove as much of the patient's outer clothing as possible.

② Wrap the patient with a cold, wet sheet until the temperature falls to about 39℃ of rectal temperature. Keep the sheet wet by pouring cold water over it continuously.

③ Fan the patient to aid in the evaporation of the water.

④ If there is no sheet available, sponge the patient with cold water.

⑤ If the patient's temperature is extremely high, immerse the patient into a tub of cold water (immersion method should be cautiously used for the unconscious patient). Careful monitoring of vital signs must be accompanied for this approach.

⑥ Place ice packs in the patient's axillae, groin, forehead, against the neck, under the arms, where large blood vessels lie close to the surface of the skin. Change the placing of ice packs in regular intervals to prevent cold injury. When the body temperature has decreased below 38℃, stop all cooling measures.

⑦ Drink some saline ice water or commercial isotonic sports drinks.

(4) Prevent Heatstroke.

① Avoid working and playing sports at the high temperatures and wet environments for a prolonged time.

② Wear light-weight, light-colored, loose-fitting clothes, and a wide-brimmed hat. Use a sunscreen with a sun protection factor (SPF) of 30 or more.

③ Some people are at increased risk for heatstroke, such as the elderly, pregnant women, people with high blood pressure, cardiovascular disease, sweating dysfunction. For these people, try to schedule exercise or physical labor for cooler parts of the day, such as early morning or evening.

④ Drink plenty of fluids, staying hydrated.

⑤ Take it easy during the hottest parts of the day.

吴山天风

CHAPTER

16/触电

Electrical Injury

电击伤，俗称“触电”，指一定强度的电流通过人体时引起的全身性或局部性组织损伤与功能障碍，重者可发生心跳和呼吸骤停，超过1000伏的高压电还可引起灼伤。闪电损伤（雷击）属于高压电损伤范畴。

16.1 临床表现

❶ 全身症状

轻者表现为精神紧张、脸色苍白、表情呆滞、呼吸心跳增快，接触部位肌肉收缩，且有头晕、心动过速和全身乏力的症状，一般很快可恢复；重者出现昏迷、持续抽搐、心跳和呼吸停止等症状。有些严重电击伤者当时症状虽不重，但在1小时后会突然恶化；有些伤者触电后，心跳和呼吸极其微弱，甚至暂时停止，处于“假死状态”，因此要认真鉴别，不可轻易放弃对触电伤者的抢救。

❷ 局部症状

（1）低压电引起的损伤：常见于电流进入点与流出点，伤面小，直径0.5—2厘米，呈椭圆形或圆形，焦黄或灰白色，干燥，边缘整齐，与健康皮肤分界清楚。一般不损伤内脏，致残率低。

（2）高压电引起的损伤：常有一处进口和多处出口，伤面不大，但可深达肌肉、神经、血管，甚至骨骼，有“口小底大，外浅内深”的特征。

16.2 预防与救治

注意生活中各种预防警告标识，提高安全意识，预防触电。

普及科学常识，如切勿用湿手、汗手接触插座电源，如需接触则需要佩戴绝缘手套；禁止在高压线缆底下钓鱼；定期检查更换家里老化的电源电器等。

救护原则为迅速将伤者脱离电源，分秒必争，尽快进行有效抢救。

1 脱离电源

迅速使伤者脱离电源，最稳妥的方法为立即关闭电闸，切断电源，若电源开关离现场太远或仓促时找不到电源开关，则应用干燥的木器、竹竿、扁担、橡胶制器、塑料制品等不导电物品将伤者与电线或电器分开。

2 轻型触电者

就地休息观察1—2小时，以减轻心脏负担，促进恢复减少意外。

3 重型触电者

对于无意识、呼吸心跳停止者必须立即进行心肺复苏，尽可能早期进行电除颤。遭受雷、电击的伤者没有心肺基础疾病，立即实施CPR，存活可能性较大，故不能轻易终止复苏。

4 创面处理

在现场应保护好伤者的电烧伤创面，防止感染，可用清洁敷料或衣服包裹。

Electrical Injury

An electrical injury is damage to the skin or internal organs when an electric current flows through a person's body. Large currents can cause fibrillation of the heart and damage to tissues, leading to general or localized tissue injury and dysfunction, or even breath and cardiac arrest.

1. Signs and symptoms

(1) Systematic symptoms.

① Less serious cases: nervousness, pale expression, changes in alertness (consciousness), irregular heartbeat, muscle spasms, pain, headache, and weakness.

② Serious cases: coma, seizures, breath and cardiac arrest. Even if the symptoms are not too serious in the beginning, some patients' conditions can deteriorate in an hour.

(2) Localized symptoms.

① Injury caused by low voltage electricity: it has one entry and one exit wound, both with small surface areas, 0.5～2cm in diameter, in the shape of a circle or an oval, brown or greyish white in color, has clear edges and is dry. Internal organs are not always injured and has a low disability rate.

② Injury caused by high voltage electricity: it has one entry and several exit wounds, which has a small surface area but are deep within the body, as muscles, nerves, blood vessels and bones could be injured massively. It characteristically has the shape of "small surface but large bottom, and shallow outside but deep inside".

2. Prevention

It is crucial to pay attention to the warning signs of the electricity in your daily life. Awareness should be taken to use electricity correctly to prevent electrical injury.

***Common sense about electricity should be popularized*:**

(1) Never mix electricity with water. You should never touch a power outlet or power source with wet hands, and wear electric insulation gloves in the case of touching them.

(2) Do not fish under a high voltage cable.

(3) Check and change electric appliances that are aging or broken.

3. Rescue

The principle for rescue is to separate the patients from the electricity ASAP.

(1) Break the contact with the electricity: Turn off the power source immediately. If the power source is too far away or cannot be found immediately, use dry wood or other non-conductive items to break the contact between the patient and electrical supply.

(2) For less severely injured patients: Rest in place for 1~2 hours to reduce the workload of the heart, promote recovery and prevent further accidents.

(3) For severely injured patients: For patients absent of consciousness, breath and heartbeat, start CPR immediately and have early defibrillation. For the patients without existing lung or heart disease, early CPR increases the chance of survival and should not be terminated in the early phase.

(4) Wound management: Protect the wound with a clean dressing or cloth to prevent infection.

阮墩环碧

CHAPTER

17 溺水

Drowning

淹溺是指人淹没于水或其他液体中，由于液体或其他杂物（水、泥沙、杂草）等堵塞呼吸道或喉头、气管发生反射性痉挛引起窒息、缺氧造成血流动力学及血液生化改变的状态。淹溺多发生在青少年、儿童及老年人，常因不慎落水、且无游泳自救能力；也可发生于企图自杀者；意外事故以洪水灾害、轮船沉没、水下作业、突发心脑血管疾病、癫痫、体育运动时防护运动设备故障或违反操作规程等为多见。严重者如抢救不及时可导致呼吸、心跳停止而死亡。现场救护是否及时是淹溺者抢救成功与否的关键。

17.1 救护方法

17.1.1 落水自救

❶ 不会游泳者

（1）被洪水卷入或落水后，应保持冷静，不要心慌意乱，要保持大脑的清醒，并冷静地将口、鼻露出水面，此时就能呼吸和呼救。

（2）当身体上浮时采取头向后仰，面向天空的姿势，先将口鼻露出水面立即进行呼吸，呼吸要浅，吸气宜深，如此反复，坚持到救援人员到来。

（3）切记不能将手上举或胡乱挣扎划水。

（4）一旦身体停止下沉并上浮时，屏住呼吸，放松全身，人体比重比水略轻，可浮出

水面。

（5）水母漂是溺水者自救方式之一。其主要做法是：头缩，双手抱膝，膝靠着胸，就可以像水母一样漂起来。要换气时，双脚慢踢，双手向前，头抬起来换气（嘴巴吸气、鼻子呼气），换完气后再缩回去。

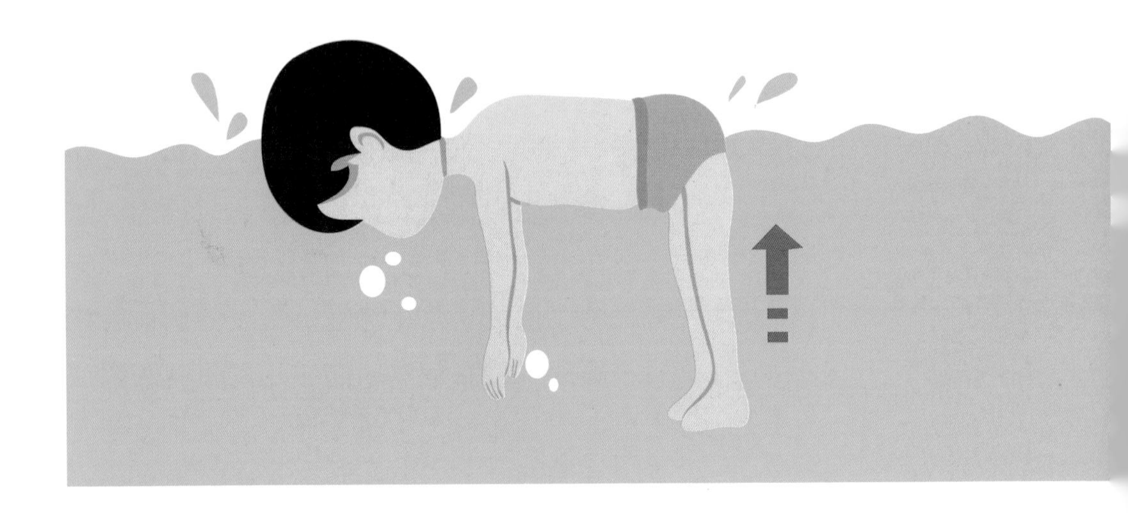

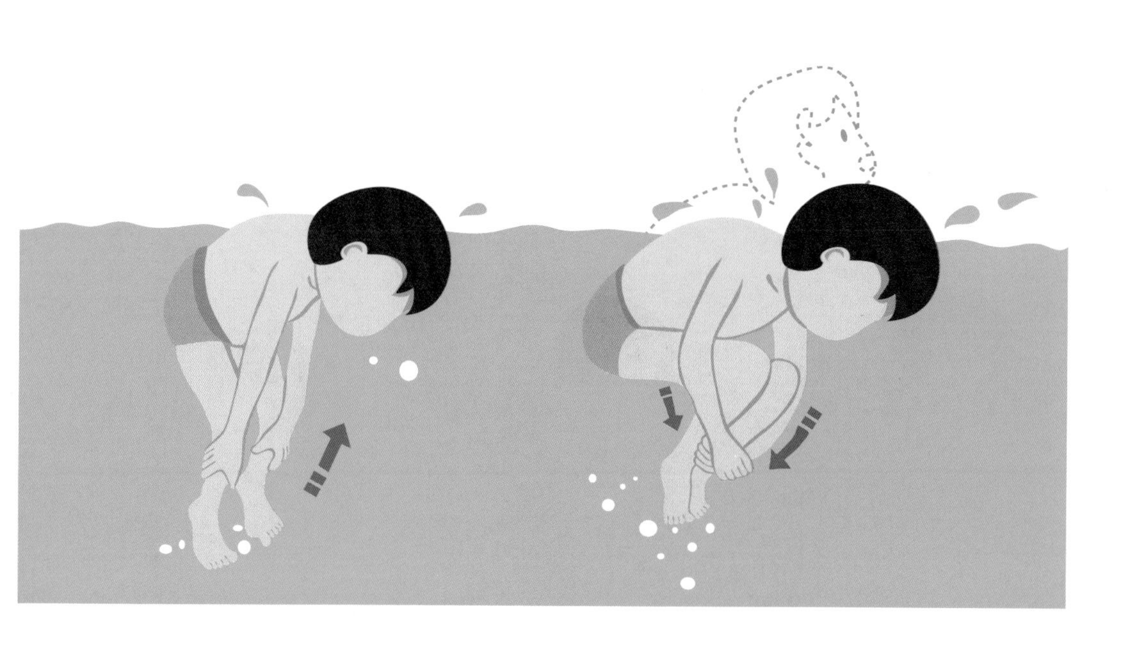

② 发生抽筋的会游泳者

会游泳的人发生溺水最大的原因在于手脚抽筋或水草缠身。

自己将身体抱成一团，脸浸入水中，抓住自己抽筋的脚，用力将大拇趾向前上方（朝向自己的方向）翘起来，持续用力，剧痛缓解后，抽筋自然就停止，然后慢慢游向岸边以防再次抽筋。若手腕肌肉抽筋者，自己将手指上下屈伸，并采取仰泳。

③ 应尽量抓住漂浮物如木板、树木等，以助漂浮，双脚像踏自行车那样踩水，并用双手不断划水。

17.1.2 水上救助

（1）水上救助，不能单凭一颗爱心就盲目下水救人，前提是必须懂水性，也就是会游泳。

（2）不会游泳者，使用如木板、树木、救生杆、绳子、救生圈（球）、救生衣、塑料泡沫等一切就近能取到的物品进行施救。注意不能用捅、打等方式，应将物品递送到溺水者身边。

最好能抓住固定物。

用脚抵住固定石块，作为支撑点。

抛掷时一定要准确到位，救护者用手紧握或用脚踩住绳子一端。

在游泳池时，大人可利用泳池扶手、固定栏杆等。

不建议小朋友采用此方法。

（3）会游泳者下水救人时，观察清楚位置。切记不可迎面救助，因为此时溺水者处于极度紧张状态，会紧紧抓住你，有可能发生一起被淹的危险。

① 对筋疲力尽的溺水者：

② 对神志清醒的溺水者：

救护者游到溺水者背后，用手从背后抱住其腋窝和胸下部或抓住溺水者手臂，拖游直至岸边。

也可双手在溺水者背后，抓住两侧腋窝，采用仰泳的姿势将溺水者拖向岸边。

③ 在水中发现淹溺者已昏迷，可在拖泳过程中向淹溺者进行口对口吹气，边游边吹，争取抢救时间。

17.1.3 岸上急救

（1）迅速将溺水者从水中拖出，设法用手指抠出其嘴里、鼻中的污泥、杂草或呕吐物，保持气道畅通。同时让周围的人拨打“120”急救电话，如身旁无人，拨通“120”至免提状态，边急救边呼救。

（2）抓紧时间做短时间倒水。

将溺水者放置在大腿上，腰部垫高、头部朝下，救助者用手按压其背部。

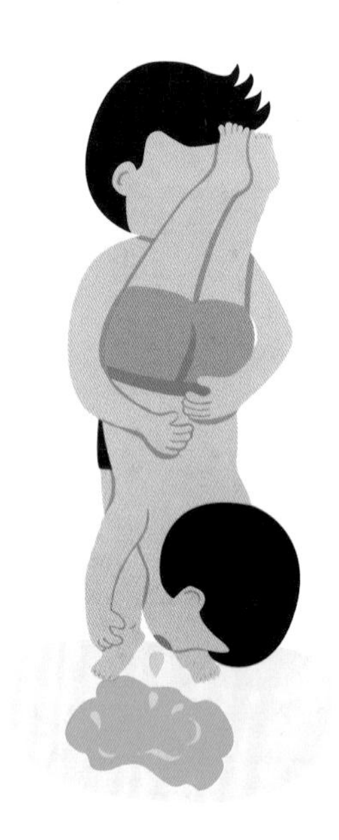

抱住溺水者腰部，使其整个人成倒立姿势。

（3）若溺水者苏醒，擦拭其身上的水分，并脱去其身上的衣服，利用身边衣物为其保暖，尤其是脚掌部位，让其安静躺着，等待救护人员到来。

（4）发现溺水者呼吸心跳停止后，立即进行心肺复苏，坚持到救护人员到来送其至医院治疗为止。

科学自救与下水救人，但愿你能了解到以上相关常识与方法，为生命赢得机会。

Drowning

Drowning is a common accident when swimming in summer and during flood period.

Rescuing method

1. Rescuing by oneself

(1) People who can't swim.

① If you are in water, keep calm and shout for help.

② Once the body stops sinking and starts to float, hold your breath and relax the whole body.

③ Jellyfish drifting: Firstly, shrink head, embrace knees with hands, with knees against the chest, and then the body will float like a jellyfish. Secondly, when taking a breath, feet kick backward slowly, and hands stretch forward, with head raises up for breath (inspiration with mouth and expiration with nose). Then repeat the first and second step.

(2) Swimmers who happen to cramp.

① Hold your body to form a clique, while face immerses into the water. Grab your cramping foot and raise the thumb forward until sharp pain reliefs. Then cramp will cease.

② If your muscles in wrist cramps, you can bend and stretch your fingers and swim by backstroke.

2. Rescuing in water

(1) If the patient is exhausted, rescuing personnel can close him from the front.
(2) If the patient is conscious, rescuing personnel can swim behind him, embrace his armpits and lower thoracic from the back of the patient with hands or seize his arms, then swim to the shore dragging the patient.

3. First aid on the shore

(1) The patient should be removed from the water as soon as possible. Try to clear mouth and nose of sludge, weeds or vomit and keep airway open. At the same time, let other people call 120 or other emergency number for help.
(2) Quickly expel out the water inhaled in the airway and stomach.
① Put the patient's abdomen across the rescuer's bent leg, keep the head down and press against the patient's back to push out the water.
② The rescuer holds the patient's waist and tips the patient upside-down.
(3) If the patient wakes up, the rescuer should wipe off the water on the body, take off the wet clothes and warm the patient.
(4) If the patient's breathing and pulse are absent, the rescuer should perform CPR immediately.

参考文献

[1] 许虹.急救护理学[M]. 北京：人民卫生出版社，2012.

[2] 浙江省红十字会. 应急救护培训手册[M]. 杭州：浙江科学技术出版社，2016.

[3] St. John Ambulance,St Andrew's First Aid and the British Red Cross. First aid manual [M]. 10th edition. London: Dorling Kindersley Limited, 2014.

[4] 香港急救暨灾难医疗培训学会，北京急救中心. 心肺复苏与创伤救护现场急救课程 [M]. 杭州：浙江科学技术出版社，2016.

[5] 陈灏珠，林果为，王吉耀. 实用内科学[M]. 北京：人民卫生出版社，2014.

[6] 陈庚生，龙绍华. 蛇伤急救与诊治[M]. 长沙：中南大学出版社，2011.

[7] 国家电网公司企业管理协会湖南分会，湖南省店里行业协会职业安全卫生分会，湖南省店里公司安全监察部. 现场触电急救知识[M]. 2版. 北京：中国电力出版社，2010.

[8] 中国疾病预防控制中心. 公众高温中暑预防与紧急处理指南[J]. 健康向导，2014，(4):48-51.

[9] 全军重症医学专业委员会. 热射病规范化诊断与治疗专家共识(草案)[J]. 解放军医学杂志，2015，40 (1):1-7.

[10] 高恒波,石汉文,田英平. 2014 年急性酒精中毒诊治共识解读[J]. 临床误诊误治，2014，27(10):5-6.

[11] 急性酒精中毒诊治共识专家组. 急性酒精中毒诊治共识[J]. 中华急诊医学杂志，2014，23 (2):135-138.

[12]中华人民共和国建设部和中华人民共和国国家质量监督检验检疫总局联合发布.中华人民共和国国家标准GB50140—2005.《建筑灭火器配置设计规范》，2005年07月15日发布,2005年10月1日实施.

[13] 中国气象局. 台风预警信号.
http://www.cma.gov.cn/2011qxfw/2011qyjxx/2011qyjxh/201110/t20111026_119423.html，2016年7月21日.

[14] WikiHow, How to Do CPR on an Adult. [online]Available at <http://www.wikihow.com/Do-CPR-on-an-Adult>(2016/06/21).

[15] WikiHow, How to Use a Defibrillator: 11 Steps. [online]Available at <http://www.wikihow.com/Use-a-Defibrillator>(2016/06/21).

[16] Mayo Clinic, Severe bleeding: First aid. [online]Available at <http://www.mayoclinic.org/first-aid/first-aid-severe-bleeding/basics/art-20056661>(2016/06/21).

[17] Professional Education, Testing and Certification Organization International (PEOI), Applying bandages and binders. [online]Available at <http://www.peoi.org/Courses/Coursesen/nursepractice/ch/ch10g.html>(2016/06/21).

[18] Medline Plus, U.S. National Library of Medicine, Snake bites. [online]Available at <https://www.nlm.nih.gov/medlineplus/ency/article/000031.htm> (2016/06/21).

[19] Mayo Clinic, Snakebites: First aid. [online] Available at<http://www.mayoclinic.org/first-aid/first-aid-snake-bites/basics/art-20056681>(2016/06/21).

[20] Medscape. Helman,R.S &Habal,R., Heatstroke Treatment & Management. [online]Available at<http://emedicine.medscape.com/article/166320-treatment> (2016/06/21).

本书绘图人员

杨　斌	杭州师范大学文创学院
林子琪	杭州师范大学文创学院
王晓丽	杭州师范大学文创学院
朱丹艳	杭州师范大学文创学院
徐　鹏	杭州师范大学文创学院
王　雪	杭州师范大学文创学院
朱梓涵	杭州师范大学文创学院
王渴鑫	杭州师范大学文创学院
赵　涵	杭州师范大学文创学院
谢　元	杭州师范大学文创学院